Communications
in Computer and Information Science   1113

Commenced Publication in 2007
Founding and Former Series Editors:
Phoebe Chen, Alfredo Cuzzocrea, Xiaoyong Du, Orhun Kara, Ting Liu,
Krishna M. Sivalingam, Dominik Ślęzak, Takashi Washio, Xiaokang Yang,
and Junsong Yuan

Editorial Board Members

More information about this series at http://www.springer.com/series/7899

Robin Doss · Selwyn Piramuthu ·
Wei Zhou (Eds.)

Future Network Systems and Security

5th International Conference, FNSS 2019
Melbourne, VIC, Australia, November 27–29, 2019
Proceedings

 Springer

Editors
Robin Doss ⓘ
Deakin University
Burwood, VIC, Australia

Selwyn Piramuthu
University of Florida
Gainesville, FL, USA

Wei Zhou ⓘ
Information & Operations Management
ESCP Europe
Paris, France

ISSN 1865-0929   ISSN 1865-0937   (electronic)
Communications in Computer and Information Science
ISBN 978-3-030-34352-1   ISBN 978-3-030-34353-8   (eBook)
https://doi.org/10.1007/978-3-030-34353-8

This Springer imprint is published by the registered company Springer Nature Switzerland AG
The registered company address is: Gewerbestrasse 11, 6330 Cham, Switzerland

Preface

Welcome to the proceedings of the Future Network Systems and Security Conference (FNSS 2019) held in Melbourne, Australia!

The network of the future is envisioned as an effective, intelligent, adaptive, active, and high-performance Internet that can enable applications ranging from smart cities to tsunami monitoring. The network of the future will be a network of billions or trillions of entities (devices, machines, things, vehicles) communicating seamlessly with one another and it is rapidly gaining global attention from academia, industry, and government. The main aim of the FNSS conference series is to provide a forum that brings together researchers from academia, practitioners from industry, standardization bodies, and government to meet and exchange ideas on recent research and future directions for the evolution of the future Internet. The technical discussions were focused on the technology, communications, systems, and security aspects of relevance to the network of the future.

We received paper submissions by researchers from around the world including the USA, Australia, New Zealand, Brazil, India, UAE, among others. After a rigorous review process that involved each paper being single-blind reviewed by at least 3 members of the Technical Program Committee, 16 full papers and 2 short papers were accepted covering a wide range of topics on systems, architectures, security, and applications of future networks. The diligent work of the Technical Program Committee members has ensured that the accepted papers were of a very high quality and we thank them for their hard work in ensuring such an outcome.

November 2019
Robin Doss
Selwyn Piramuthu
Wei Zhou

Organization

General Chairs

Robin Doss   Deakin University, Australia
Selwyn Piramuthu   University of Florida, USA
Wei Zhou   ESCP Europe, France

Program Committee

S. Agrawal   Delhi Technological University (DTU), India
Aniruddha Bhattacharjya   Tsinghua University, China
Arcangelo Castiglione   University of Salerno, Italy
Miguel Franklin de Castro   Federal University of Ceará, Brazil
Manoj Joy   Amal Jyothi College of Engineering, India
Mounir Kellil   CEA LIST, France
Hang Li   Shenzhen Research Institute of Big Data, China
Li Liu   Chongqing University, China
Nagendra Kumar Nainar   Cisco, USA
Shashikant Patil   SVKM NMIMS, India
Hussain Saleem   University of Karachi, Pakistan
Saeid Samadidana   Austin Peay State University, USA
Koushik Sinha   Southern Illinois University, USA
Dimitrios Stratogiannis   National Technical University of Athens, Greece
Carlo Vallati   University of Pisa, Italy
Hui Wu   University of New South Wales, Australia
Keun Soo Yim   Google, Inc., USA

Contents

Emerging Networks and Applications

Intrusion Detection System Classifier for VANET Based on Pre-processing Feature Extraction

Ayoob Ayoob[1], Ghaith Khalil[2(✉)] ⓘ, Morshed Chowdhury[2] ⓘ, and Robin Doss[2] ⓘ

[1] Department of Electronics and Information Engineering,
Huazhong University of Science and Technology, Wuhan 430074,
People's Republic of China
I201522060@hust.edu.cn
[2] School of Information Technology, Deakin University,
Geelong, VIC, Australia
g.al@deakin.edu.au

Abstract. Vehicular Ad-hoc Networks (VANETs) are gaining much interest and research efforts over recent years for it offers enhanced safety and improved travel comfort. However, security threats that are either seen in the ad-hoc networks or unique to VANET present considerable challenges. In this paper, we are presenting the intrusion detection classifier for VANET base on pre-processing feature extraction. This ID infrastructure novel is mainly introducing a new design feature for extraction mechanism a pre-processing feature-based classifier. In the beginning, we will extract the traffic stream structures and vehicle location features in the VANET model. Later an Algorithm Pre-processing feature-based classifier was designed for evaluating the IDS by using hierarchy learning process. Finally, an additional two-step validation mechanism was used to determine the abnormal vehicle messages accurately. The proposed method has better finding accuracy, stability, processing efficiency, and communication load.

Keywords: Vehicular Ad-hoc Network (VANET) · Intrusion Detection System (IDS) · IEEE 802.11p

1 Introduction

Over the last years, there was a growing interest in the area of Vehicular Ad-hoc Networks (VANET). This was due to the variety of services it can award. These services can be classified to safety and non-safety applications. Safety is one of the most crucial objectives of VANET. As safety will reduces accidents, save lives and reduce traffics. Among side safety, other services such as Internet access, weather forecast, and geolocation information can enrich travel experience by providing travel comfort, convenience and infotainment. Committees are devoting efforts to finalize standards for VANET. These standards include IEEE 1609.x, 802.11p and Wireless Access in Vehicular Environment (WAVE). WAVE is a layered architecture for

© Springer Nature Switzerland AG 2019
R. Doss et al. (Eds.): FNSS 2019, CCIS 1113, pp. 3–22, 2019.
https://doi.org/10.1007/978-3-030-34353-8_1

devices complying IEEE 802.11 to operate on Dedicated Short-Range Communication (DSRC) band. The IEEE 1609 family defines the architecture and the corresponding protocol set, services and interfaces that allow all WAVE stations to interoperate within the VANET environment. Vehicle Ad Hoc Network (VANET) is one of the developing technologies that can dramatically change ways of communication and industry. It suggestively rallies the quality of human beings and will become a reality soon. Many vehicle makers began incorporating wireless access in the vehicular environment (WAVE) in their vehicles in 2015 [1] and [2]. WAVE is a technology designed on the IEEE 802.11p protocol that provides the broadcast standard for the short-range communications technology (DSRC). VANET enables wireless communication between vehicles through the DSRC including vehicle-to-vehicle communication (V2V) as well as vehicle infrastructure communication (V2I) [3] and [4]. This will allow vehicles to communicate with each other's and give drivers the option to decide. As VANET manages traffic information that are linked to human life safety, this will improve the safety of driving the vehicles which also increase the need of security in VANET. To address the security issues in VANET Intrusion Detection Systems (IDS) is deployed inside each vehicle to detect internal and external security threats and network attacks such as Blackhole attack, Denial of Service (DoS), Sybil attack, etc.

2   A Literature Review on Vehicle Ad Hoc Network (VANET) Models

Analyzing the VANET information is one of the effective ways to protect VANET and identifying unusual activities in the VANET then alerting it. In [5] the authors propose a novel framework with preservation and repudiation (ACPN) for VANETs. In ACPN, they introduce the public-key cryptography (PKC) to the pseudonym generation, by obtaining vehicles' real IDs. As they stated that the self-generated PKC based pseudonyms are also used as identifiers instead of vehicle IDs, while the update of the pseudonyms depends on vehicular demands.

In [6, 7] and [8] the authors propose a system to detect attacks on vehicles based on matching IDS. It has high detection accuracy and efficiency, that can only detect attacks that match their rules and ignore other unknown attacks. In [9], a statistical-based approach was used for IDS. In [7], a new method was Introduced in IDS which uses Bayesian theory to switch IDS status (active or idle) to reduce the detection time. Yet, it is hard to detect attacks that occur during the IDS's idle state, which can compromise the accuracy of IDS detection.

While the authors in [4] proposed a security framework for VANETs to achieve privacy desired by vehicles and non-repudiation required by authorities, in addition to satisfying fundamental security requirements including authentication, message integrity and confidentiality. They stated that the proposed framework employs an ID-based cryptosystem where certificates are not needed for authentication as it increases the communication efficiency for VANET applications where the real-time

constraint on message delivery should be guaranteed. They also briefly review the requirements for VANET security and verify the fulfillment of our proposed framework against these requirements.

The data-centric trusty mechanism main function is to create a message processing center, which evaluates and broadcasts all the messages [6, 8] and [4]. In [6], the data-centric trusty mechanism was used to discriminate abnormal behaviors by only considering shared data. In [4], the author proposed another VANET model to detect and correct errors for the data which was sent by the vehicle While in [8], the emergency message is relayed, and fake information are identified according to the message type and the subsequent behaviors of the vehicle sender. In [9] it proposed an anomaly detection system and eviction mechanism. The vehicle was considered as an abnormal behavior if the message sent by the vehicle did not accord with the general situation. Once the vehicle is classified as attacking vehicle, the neighbor's vehicle can temporarily deport it by sharing the warning message. Unlike the previous two technologies, IDS, as a solution to VANET security issues, effectively detects attacks by analyzing and classifying the messages in the VANET. It accurately identifies known attacks and attempts to predict new types of attacks and protect the system from unknown attacks as well. Yet, in this intrusion prediction system, the probability threshold needs to be set sensitively to obtain accurate results. Some severe security issues need to be resolved before using IDS to prevent attacks in VANET. A number of other studies have been conducted in the IDS area of VANET as in [10] and [11], where framework that combines creditworthiness scores with rule-based detection is used for IDS in VANET. However, there the particularly serious disadvantage of false alarms, detection times and load problems when the number of vehicles increases.

In [12] the authors stated that the Vehicular Ad-hoc Networks (VANETs) are vulnerable to various type of network attacks like Blackhole attack, Denial of Service (DoS), Sybil attack etc., and they proposed an Intrusion Detection Systems (IDSs) to address these security threats also they stated that VANETs operate in bandwidth constrained wireless radio spectrum. They introduced a IDS framework to represent a significant volume of IDS traffic are not suitable for VANETs. Also, they stated that dynamic network topology, communication overhead and scalability to higher vehicular density are some other issues that needs to be addressed while developing an IDS framework for VANETs. Their research focus was to address the above issues by proposing a multi-layered game theory-based intrusion detection framework and a novel clustering algorithm for VANET. They claimed that the communication overhead of the IDS is reduced by using a set of specification rules and a lightweight neural network-based classifier module for detecting malicious vehicles. Also, they claimed that the volume of IDS traffic is minimized by modelling the interaction between the IDS and the malicious vehicle as a two player non-cooperative game and adopting a probabilistic IDS monitoring strategy based on the Nash Equilibrium of the game.

For addressing VANET security issues, there are lots of different solutions which includes creditworthiness mechanisms [13] and [14] and data-centric trusty mechanisms [15]. The creditworthiness mechanism is a method based on the idea which gives

the vehicle scores according to their historical behaviors and current behaviors so that a vehicle has a higher score of safety with higher ratings and a vehicle is regarded as a suspected vehicle with lower scores. In other words, we can state that the creditworthiness mechanism [10] is giving trust scores to vehicles in according to its historical or current interaction, where the trust score indicates the creditworthiness of the vehicle in VANET. According to scores, the vehicle behaviors are restricted to different degrees. There are various creditworthiness frameworks which were presented in [10] and [14]. In [14], the authors adopted a decentralized infrastructure to deploy a creditworthiness mechanism. A centralized infrastructure was proposed to implement a reputation mechanism. While in [10] the authors proposed a novel ID-based authentication framework with adaptive privacy preservation for VANETs. In this framework, adaptive self-generated pseudonyms were used as identifiers instead of real-world IDs. The authors explained that an update of the pseudonyms depends on vehicular demands proposed. And the ID-Based Signature (IBS) scheme and the ID-Based Online/Offline Signature (IBOOS) scheme were used, for authentication between the Road Side Units (RSUs) and vehicles, as well as authentication among vehicles, respectively. Also, a system evaluation has been executed using efficient IBS and IBOOS schemes. They claimed that it shows that, the proposed authentication framework with privacy preservation was suitable to the VANET environment.

3   Vehicle Ad Hoc Network (VANET) Model

A generic VANET structure can be obtained according to the previous VANET [16], as shown in Fig. 1 [24]. In this VANET structure, each vehicle will be installed several devices, such as: GPS, Radar, IDS and so on. Among them, GPS can get the current position of the vehicle. Radar is used to measure the communication signals among vehicles. IDS are used to detect attacks from inside and outside the vehicle. By sending a message, the vehicle can communicate with each other vehicles within its communication domain. At the moment of vehicle communication, vehicles can be divided into three different roles (current vehicle, neighbor vehicle and target vehicle). The current vehicle is defined for each vehicle itself, and the neighbor vehicle is the vehicle in the communication domain of the current vehicle, and the target vehicle is a special neighbor vehicle whose message is being processed by the current vehicle. In addition, each vehicle contains three data tables, namely: the historical neighbor message table, the current neighbor message table and the vehicle location information table. The data in these tables will be used in driving decisions to enhance the driving experience.

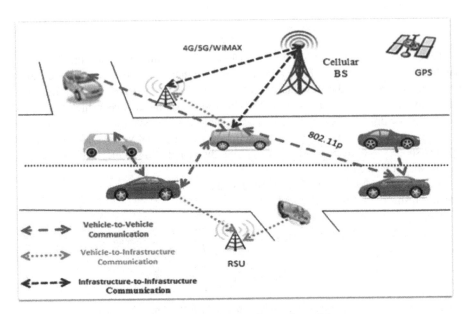

Fig. 1. Vehicle Ad Hoc Network model.

3.1   VANET Model Measurements

In a freeway, the vehicle can calculate the density of vehicles by counting the number of different vehicles in the historical neighbor's message table. In addition, each vehicle gets location coordinates (*Xposown* and *Ypsosown*) by GPS. It is important to note that the vehicle does not have direct access to the flow of traffic ($\overline{\text{Flowown}}$) and the distance between two vehicles (*Do&n*). The $\overline{\text{Flowown}}$ needs to be calculated from the model of Greenshields [17], and (*Do&n*) can be calculated from the free space model. The Greenshield model is considered as a fairly accurate model for describing the relationships among speed (*v*) (Km/hr), density (*k*) (veh/Km) and traffic flow (*q*). The parameter v_f can be defined as the free mean speed at which the vehicle density is zero. With the density increasing, the speed will decrease, until the maximum k_j (congestion density) (veh/Km) is achieved. This is the case of vehicle congestion. The relationship between speed and density is expressed as follows:

$$v = v_j - \frac{k}{k_j} v_j \tag{1}$$

$$q = k \times v \tag{2}$$

Hence,

$$q = v_j k \times v \frac{k^2}{k_j} \tag{3}$$

$$q_m = v_m k_m \tag{4}$$

Greenshields model shows the relationship between traffic flow and the density is parabolic curve. When flow is very low, speed is higher. The drivers are able to travel at a desired speed. As the flow increases, speed gradually decreases. The highest flow shows the transition of non-congested to congested condition. Greenshields's model shows the relationship between speed and density as follows:

where k_m and v_m parameters can be defined by the optimal vehicle density and the optimal speed value, respectively. In this point, the vehicle flow achieves the maximum q_m allowed (Figs. 2 and 3).

MaxSpeed and *MaxDensity* can be defined by free flow speed and congestion density, respectively in Eqs. (5 and 6):

$$\overline{Speedown} = MaxSpeed - \frac{Density_{own}}{MaxDensity} MaxSpeed \tag{5}$$

$$Flow_{own} = Density_{own} \times \overline{Speedown} \tag{6}$$

From the equations above we can produce the following formulas:

$$\overline{Speedown} = MaxSpeed - \frac{MaxSpeed}{MaxDensity} Density_{own} = v_f - \frac{v_f}{k_j} k \tag{7}$$

$$Flow_{own} = Density_{own} \times \overline{Speedown} = k * \left(v_f - \frac{v_f}{k_j} k\right) \tag{8}$$

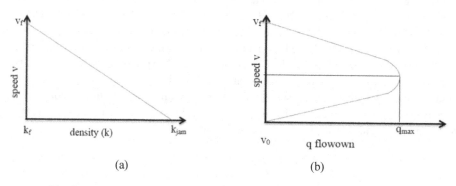

(a)   (b)

Fig. 2. Explains the graphical relation between the density, speed and flow.

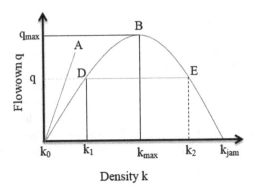

Fig. 3. (a) Speed vs density, (b) Speed vs flowown and (c) Flowown vs density

The free space model was adopted in order to define Do&n model, where $Pr(d)$ is the transmit power, d which is the distance between sender and receiver. Pt is the transmit signal strength, and τ is the wave length. In this model. At this point, the parameter d is the only variable. If the value of d is determined, then the model can be uniquely identified.

$$r\,(d) = \frac{\tau Pt}{(4\pi)^2 (d)^2} \tag{9}$$

Since each vehicle is equipped with radar, when it receives a message from a neighbor's vehicle, it receives the signal strength of neighbor's vehicle ($RSSIneg$), the wavelength ($WLneg$) and transmits power ($SPneg$). Based on the free space model, the vehicle can calculate Do&n according Eq. (10).

$$Do\&n = \left(\frac{SPneg \times WLneg^2}{(4\pi)^2 RSSIneg} \right)^{\frac{1}{2}} \tag{10}$$

3.2 VANET Message Format

In order to communicate with the neighboring vehicles in the current vehicle's communication domain, each vehicle broadcasts a beacon message (*BeaconMsg*) for a fixed time period (*BeaconT*) in the format shown in Eq. (11). Where *IDoW* is the current vehicle ID.

$$BeaconMsg\left(IDown, \overline{Flowown}, X_{posown}, Y_{posown}\right) \qquad (11)$$

When the current vehicle needs to get the information of the target vehicle from its neighbor vehicle, it will broadcast request message in the format of request message (IDown, IDtag), where IDtag is the ID of the target vehicle. Once the neighbor's vehicle receives request message, it will broadcast their location coordinates X_{posown}, Y_{posown} and their distance from the target vehicle ($Dn\&t$) through $ResroonseMsg$. $ResroonseMsg$ format which can be defined by $responseMsg(IDneg, IDtag, X_{posown}, Y_{posown}, Dn\&t)$. The duration of this process is called response waiting time ($WaitingT$).

4   VANET Information

In order to store the information in VANET (Table 1), each vehicle has three tables. Those tables are the current neighbor message table, the historical neighbor message table and the location information table respectively. The current neighbor message table and the historical neighbor message table are used to store $BeaconMsg$ from a neighbor's vehicle. The location information table is used to store $ResponseMsg$. Each of these three tables has an update cycle, which is the same as for $BeaconT$. When the update time is up, all the contents in the location information table and the historical neighbor message table are deleted, and the content of the current neighbor message table is transferred to the historical neighbor message table.

Table 1. VANET information table

Current neighbor message table	Store $BeaconMsg$
Historical neighbor message table	Store $BeaconMsg$
Location information table	Store $ResponseMsg$

4.1   Proposed Method

In order to train the IDS conditions), so that the IDS can detect the biased VANET message according to presented in this chapter, the training process needs to be performed in a non-attack vehicle situation (under normal the normal model (as shown in Fig. 4).

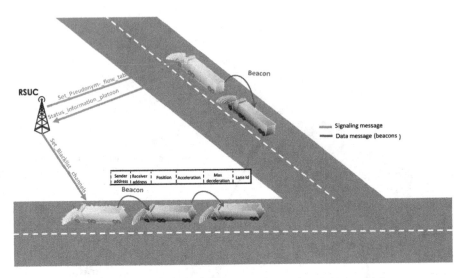

Fig. 4. A vehicle receives beacon messages from neighbor vehicles.

4.2 Pre-processing Feature Extraction

As shown in Fig. 5 the algorithm has two main parts. The first part is the calculation of traffic flow characteristics *FlowR*. It is based on the principle that the traffic volume of each vehicle should be very similar to the traffic volume of its neighboring vehicles under the same traffic conditions. If an attacking vehicle wants to create a non-existent incident by sending its reduced fake traffic to other vehicles, its *FlowR* will be different from normal. Therefore, false information attacks can be detected by *FlowR*. The second part is to calculate the vehicle position feature, PositionR. PositionR is the deviation between the position coordinate of the neighboring vehicle declaration and the measurement position. When a neighboring vehicle sends a fake location message, it is declaring the location coordinates will go beyond the normal deviation range. Therefore, with PositonR, not only can detect false message attacks, but also detect what attacks.

Although vehicle location can be verified by any VANET observation model such as Accepted Signal Strength (RSS), Time of Arrival and Time of Arrival. Taking into account the relatively easy access to RSS, this program uses RSS measurement. In the semi-cooperative detection method, the vehicle position feature can be calculated by either of the following two cases.

If *IDtag* does not exist in the historical neighbor table (cooperation case), each vehicle first broadcasts request message to its neighbor's vehicle. When its neighbor's vehicle receives the *RequestMsg*, they will broadcast *ResponseMsg* if they did not receive the *IDtag* for the *RequestMsg* in *WatingT*. When the time reaches *WatingT*, the current vehicle populates *ResroonseMsg* of each neighboring vehicle in its position

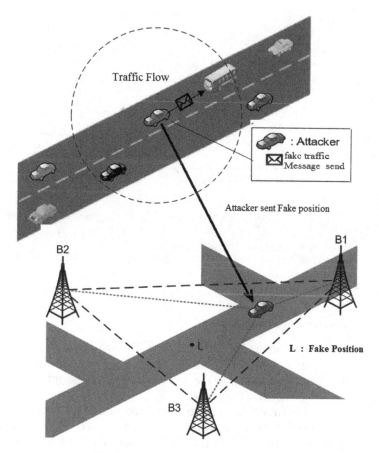

Fig. 5. Preprocessing Feature Extraction

information table. Then, each vehicle calculates $Bias_{o\&t}$ and $Bias_{n\&t}$. The former is the position deviation between itself and the target vehicle, as shown in Eq. (12). The latter is the deviation between the neighbor's vehicle and the target vehicle, as shown in Eq. (13) (there may be more than one $BiaSn\&t$). Next, each vehicle checks $|Bias_{o\&t} - Bias_{n\&t}| < MaxGap$ for $ResponseMsg$ from the attacker vehicle, where $MaxGap$ was determined during training, which is the largest difference between $Bias_{o\&t}$ and $BiaSn\&t$.

$$Bias_{o\&t} = \left| D_{o\&t} - \sqrt{\left(Xpos_{own} - Xpos_{tag}\right)^2 + \left(Ypos_{own} - Ypos_{tag}\right)^2} \right| \quad (12)$$

$$Bias_{n\&t} = \left| D_{n\&t} - \sqrt{\left(Xpos_{neg} - Xpos_{tag}\right)^2 + \left(Ypos_{neg} - Ypos_{tag}\right)^2} \right| \quad (13)$$

Finally, if Bias ≥ 3, PositionR is obtained from Eq. (14). Otherwise, the characteristic *PositionR* can not be obtained.

$$\text{PositionR} = \frac{1}{n}\left(\sum Bias_{o\&t} + Bias_{n\&t}\right) \tag{14}$$

If there is an *IDtag* (non-cooperation) in the neighbor table, *PositionR* of the current vehicle can be directly obtained without the help of a neighbor's vehicle as shown in Eq. (19). This means that the current vehicle does not need to send *RequestMsg* and wait for *ResponseMsg*. In Eq. (19), *Biast t*∗ is the position deviation between the current target vehicle and its previous moment, which can be obtained from Eq. (16). *Biaso ∗ &t* is the position deviation between itself and the previous phase of the current target vehicle, as can be seen from Eq. (18), Which shows a method of calculating the distance (*Dt&t* ∗) between the historical position of the target vehicle and the current position of the target vehicle and the distance (*Do ∗ &t*) between the current position of the current vehicle and the current position of the target vehicle. Because *IDtags* exist in the historical neighbor table, each vehicle has both historical and current information of its own vehicle and the target vehicle. Therefore, *Dt&t* ∗ can be obtained from Eq. (15), where (*Dt&t* ∗ is closest to the value of *Do&o* ∗). *Do ∗ &t* can be obtained from Eq. (17) according whether *Dt&t* ∗ and *Do&o* ∗ already obtained.

$$D_{t\&t'} = \left(\frac{\text{MaxSpeed}}{2} \pm \sqrt{\frac{\text{MaxSpeed}}{2} - \frac{\text{MaxSpeed}}{2} \text{AvgFlow}_{tag}}\right) \times \text{Beacont} \tag{15}$$

$$\text{Bias}_{t\&t} = \left| D_{t\&t'} - \sqrt{\left(\text{Xpos}_{tag} - \text{Xpos}_{tag'}\right)^2 + \left(\text{Ypos}_{tag} - \text{Ypos}_{tag'}\right)^2} \right| \tag{16}$$

$$D_{o'\&t} = \sqrt{D_{o\&t}^2 + D_{o\&o'}^2 - 2D_{o\&t}D_{o\&o'}\cos < TOO'} \tag{17}$$

$$\text{Biaso} \ast \&t = \left| Do\&t - \sqrt{(\text{Xposown*} - \text{Xpostag})^2 + (\text{Yposown*} - \text{Ypostag})^2} \right| \tag{18}$$

$$\text{PositionR} = \frac{Biaso\&t + Biast\&t \ast + Biaso \ast \&t}{3} \tag{19}$$

Algorithm 1. Algorithm for IDS

---

for each message receive do
20 -   Update *density flow*
         Update speed Flow
         Update *position R*

*Flowown = Speed Flow * density Flow*
if *Floweng – Flowown ≤ Threshold Then*

   Calculate Flowown
   Accepted Data then go to **10**

   else

      Detection Attack
      Rejected Data
10-   Calculate devition between current Vehicle & target Vehicle
         calculate *Biaso &1*
         calculate *Biasn &1*
   if *Bias o&t-Biasn&t| < MaxGap*

calculate *position R (* Vehicle Position Feature)
Accepted Data go to **20**

   else

Position R not obtain (detection Attack)
Detection Attack
Rejected Data

         end f
      end f
end for

---

5   Performance Evaluations

5.1   Experimental Environment

Simulation experiments are based on NS2 [18, 19] and urban traffic simulation tools (SUMO) [20]. NS2 provides a relatively complete low-level protocol and a simple

programming interface. SUMO is a software tool used to generate vehicle traffic by specifying the speed, type, behavior, and number of vehicles. SUMO can also set the road type and conditions. While simulating the IDS in this article, SUMO is used to generate the move trace file, and NS2 is used to load these trace files and run the IDS.

Table 2. Parameter Setting.

Parameter	Value
Experimental scene	2 lane of 5 km highway
Maximum vehicle speed	100 km/h
Wireless communication protocol	802.11p
Transmission range	500 m
Simulation time	165 s
Vehicle arrival interval	1 s
Transmission interval	0. 5 s
The corresponding waiting times	0. 2 s
Tau1, Tau2	0. 1, 0, 01

In the experimental simulation, the experimental parameters are shown in Table 2. The vehicle can run on a 5 km long highway with 2 lanes and can communicate with other vehicles within 500 m of the transmission range according to the communication protocol 802.11p. To avoid generating too much data in a simulation, we set the simulation time to 165 s, the vehicle arrival interval to 1 s and the transmission interval (*BeaconT*) to 0.5. In order to ensure that *ResponseMsg* has enough time to arrive, this article sets the response latency (*WaitingT*) to 0.2 s. Tau1 and Taut (HGNG.related two parameters) were set to 0.1 and 0.01.

In order to verify the proposed IDs system in Fig. 5. The following experiment data collected uses three different simulation scenarios as shown in Table 3.

Table 3. Show a sample of collected data

Sample	Time (s)	Flowown	Location	ID
1	1	0.593235	23.72049	0
5	8	0.644053	42.12209	0
9	16	0.646086	41.05204	0
13	24	0.703003	16.85941	0
17	32	0.7152	29.9111	0
21	40	0.754839	23.90592	0
25	48	0.755855	15.90285	0
29	56	0.703003	68.40415	0
33	64	0.663026	14.943	0
37	72	0.663703	6.860127	0
41	80	0.646086	34.15423	0

(continued)

Table 3. (*continued*)

Sample	Time (s)	Flowown	Location	ID
45	88	0.654217	46.3577	0
49	96	0.654217	27.25388	0
53	104	0.654217	14.14375	0
57	112	0.63389	36.41111	0
83	1	0.618956	20.82476	1
87	8	0.639966	22.92278	1
91	16	0.66371	18.83949	1
95	24	0.710135	15.98234	1
99	32	0.729328	20.93495	1
103	40	0.802107	23.0903	1
107	48	0.761356	17.20079	1
111	56	0.7094	22.15436	1
115	64	0.670765	16.26485	1
119	72	0.670885	20.25653	1
123	80	0.651931	24.36481	1
127	88	0.664776	25.47179	1
131	96	0.66034	18.70688	1
135	104	0.661367	14.49339	1
139	112	0.641506	16.66332	1

5.2 Experimental Results

In order to verify the validity of proposed IDS, the experiment collect the average flow rate (Flowown) of vehicles randomly selected under different conditions and the flow of its reception (Floweng), as shown in (Figs. 6 and 7). Figure 6 shows the data under normal conditions. It can be observed (Flowown and (Floweng) values are fairly close with very slight variation. Then, 50% of the vehicles attacked are inserted and they send a fake traffic flow (close to the blue dot at the bottom of the Fig. 6 and 4.8) after $t = 50$ seconds. In the absence of IDS (Fig. 7), (Floweng) decreases with acceptance of all messages (including normal and abnormal messages) and decreases from the legitimate vehicle (Floweng) (the blue dot near the red curve) as it also Affected vehicles. In the case of IDS, as shown in Fig. 8, it can be observed that the proposed IDS can detect Floweng and then reject the abnormal message, so that the (Flowown) value is similar to normal.

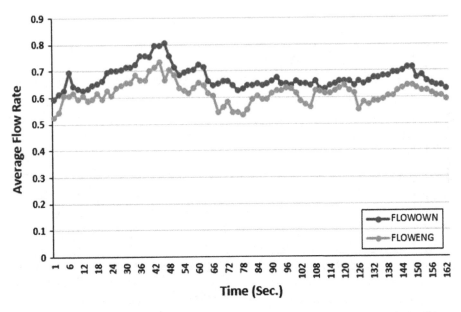

Fig. 6. Average Flow (Flowown) and Reception Flow (Floweng) Rates in Normal Conditions.

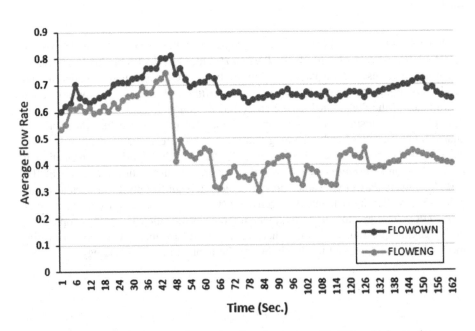

Fig. 7. Average flow rate and reception flow time rates with 50% vehicle attacks.

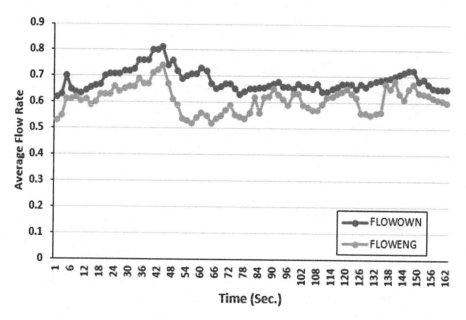

Fig. 8. Average flow rate and reception flow time rates with 50% vehicle attacks (with adopted IDS).

In verifying the validity of the IDS proposed in this paper for vehicle position detection, the experiment collects the *PositionR* values of vehicles randomly selected in different situations and displayed in Fig. 6. First run the simulation under normal conditions. Figure 8a presents the values of *PositionR* at a relatively low level. The coordinates of the PositionR here are caused by some unavoidable errors, for example device errors, transmission errors, and so on. Hence, a noticeable variation was attained. On the other hand, assuming that there is 50 percentage of attack vehicles in an abnormal situation, the attacking vehicle changes their position information in *BeaconMsg* and *ResponseMsg*, as exhibited in Fig. 9b and c. In the absence of IDS (Fig. 9b), many *PositionR* values appear to deviate significantly from normal at this point. Finally, IDS in each vehicle is activated (Fig. 9c), it can be notice here almost all *PositionR* values are close to normal, which indicates the efficiency of the adopted IDS in the vehicle position detection. Therfore and based on those two-step verification mechanism, it is approved the feasibility of the applied modeling strategy.

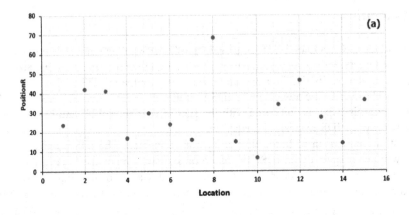

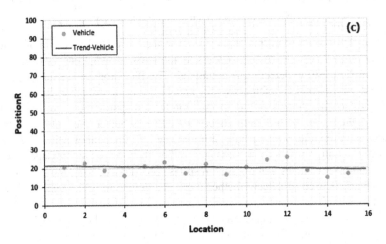

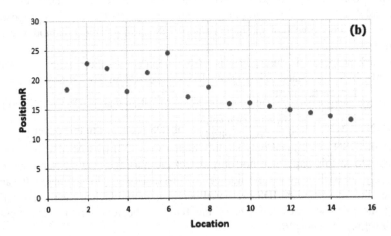

Fig. 9. (a) PositionR of Vehicles in Normal Conditions. (b) PositionR of Vehicles in Normal Conditions with 50% Vehicle Attacks (without adopted IDS). (c) PositionR of Vehicles in Normal Conditions with 50% vehicle attacks (adopted IDS).

6 Validation

In this article, we compared IDS proposed by other researchers and IDS proposed by other researchers with accuracy, stability, processing efficiency and message size. The comparison here is divided into two parts. First, we compared IDS in this paper with IDS in [21] to verify the advantage of IDS in traffic flow detection. Secondly, the IDS proposed in this paper is compared with the IDS of references [22] and [23] to verify the advantage of IDS in position detection.

First, a comparison of traffic flow tests was conducted. The IDS proposed in this paper will be compared with IDS [19] based on statistical methods. In [19], the author uses a hypothesis test to detect whether there is a deviation of the traffic flow of the attacker's vehicle. The result show as the IDS proposed in this paper is superior to IDS [19] in terms of TD which is equal $(TD = \dfrac{\text{No. of vehicles been classified correctly}}{\text{No. of total vehicles}})$ and FD which is equal $FD = \dfrac{\text{No. of vehicles been classified wrongly}}{\text{No. of total vehicles}}$. In addition, because the proposed IDS proposed in this paper does not need to collect a certain amount of traffic information before the testing process as the basis of judgment, as [19].

Next, vehicle position detection is compared. The IDS proposed in this paper will be compared with two IDS references [22] and [23]. As described in Ref. [22], the author first collects neighbor location information from response messages of neighbor vehicles and then uses information theory to detect whether the vehicle location information is biased after waiting time. In reference [23], the author collects neighbor location information from beacon messages rather than response messages (which means that neighbor vehicles will not send response broadcasts). After the beacon time, a hypothesis test is used to detect whether there is a deviation in the vehicle position information.

The IDS proposed in this paper has TD and FD values higher than IDS in references [22] and [23], and the uncertainty rate is lower than the two IDSs because GIISOM-based IDS proposes a new semi-collaborative mechanism that not only checks for positional discrepancies through collaboration but also locates the target vehicle based on historical location information. This means that the IDS proposed in this paper has higher precision and higher stability than the statistical method in vehicle position detection.

The IDS proposed in this paper is also compared with the APS and ASMN values of IDS reference [22] and [23]. The result is shown the APT $(APT = \dfrac{\sum \text{PT of each vehicle}}{\text{No. of each vehicle}})$ in this paper is smaller than that of IDS in [22] and [23]. The ASMN $ASMN = \dfrac{\sum \text{SMN of each vehicle}}{\text{No. of each vehicle}}$ of IDS proposed in this paper is smaller than that of IDSN in [22] and only slightly higher than [23] IDS ASMN, because the IDS proposed in this paper can not only detect the message in a shorter time (even immediately), but also reduce the number of messages sent. In summary. It can be concluded that the proposed IDS presented herein is more efficient than the two IDSs and that the proposed message size of the IDS can be reduced to avoid message congestion [24, 25] and [26].

7 Conclusion

In this research paper, we had proposed the intrusion detection for VANET base on pre-processing feature extraction. The processing features-based IDS used to detect attacks in the VANET. The IDS mainly comprehend new feature extraction mechanism and classifier based on Pre-processing Feature Extraction Algorithm. To obtain the location features more effectively, an extraction mechanism is introduced to extract two primary features, traffic flow features, as well as the vehicle position features. Also, we adopt a semi-cooperative way to recover the location of the vehicle. The other feature beside cooperatively collecting the current location information of the neighboring vehicles is to extract the position features according to the historical location information. The additional two-step verification mechanism which was presented to achieve more accurate judgment of vehicle message anomalies throughout the conducted experiment in which was evaluated in the Proposed IDS. In our conclusion that this proposed system was more precise and stable from other available IDS.

References

1. Wu, J., Wang, Y.: Performance study of multi-path in VANETs and their impact on routing protocols. Wirel. Eng. Technol. **2011**, 125–129 (2011)
2. Pradweap, R.V., Hansdah, R.C.: A novel RSU-aided hybrid architecture for anonymous authentication (RAHAA) in VANET. In: Bagchi, A., Ray, I. (eds.) ICISS 2013. LNCS, vol. 8303, pp. 314–328. Springer, Heidelberg (2013). https://doi.org/10.1007/978-3-642-45204-8_24
3. Al, G.: Technology Design Principle, Application and Controversies. Nov. Sci. Publ. (2018)
4. Rajarajan, C., Zaidi, K., Milojevic, M., Member, S., Rakocevic, V.: City research online city, University of London institutional repository host based intrusion detection for VANETs: a statistical approach to rogue node detection. IEEE Trans. Veh. Technol. **65**(8), 6703–6714 (2016)
5. LAN/MAN Standards Committee of the IEEE Computer Society Part 11 : Wireless LAN Medium Access Control (MAC) and Physical Layer (PHY) Specifications IEEE Computer Society, vol. 2012 (2012). ISBN 9780738172118
6. Hao, Z., Singh, G., Kamboj, E.S.: A review on multiple malicious and irrelevant packet detection in VANET. Int. J. Technol. Comput. 2 (2016)
7. Sedjelmaci, H., Senouci, S.M., Ansari, N.: Intrusion detection and ejection framework against lethal attacks in UAV-aided networks: a bayesian game-theoretic methodology. IEEE Trans. Syst. **18**, 1143–1153 (2017)
8. Erritali, M., El, O.B.: A survey on VANET intrusion detection systems. Int. J. Eng. Technol. A **5**, 1985–1989 (2013)
9. Zaidi, K., Milojevic, M.B., Member, S., Rakocevic, V.: Host-based intrusion detection for VANETs: a statistical approach to rogue node detection. IEEE Trans. Veh. Technol. **65**, 6703–6714 (2016)
10. Lu, R., Lin, X., Luan, T.H., Liang, X., Member, S., Shen, X.S.: Pseudonym changing at social spots: an effective strategy for location privacy in VANETs. IEEE Trans. Veh. Technol. **61**, 86–96 (2012)
11. Wex, P., Breuer, J., Held, A., Leinm, T.: Trust issues for vehicular ad hoc networks. Veh. Technol. Conf. **4**, 2800–2804 (2008)

12. Bibhu Vimal, K.R.: Performance analysis of black hole attack in VANET. Comput. Netw. Inf. Secur. **5**, 47–54 (2012)

13. Ayoob, A., Su, G., Al, G.: Hierarchical growing neural gas network (HGNG)-based semi cooperative feature classifier for IDS in vehicular ad hoc network (VANET). J. Sens. Actuator Netw. **7**, 41 (2018)

14. Minhas, U.F., Zhang, J.: Towards expanded trust management for agents in vehicular ad-hoc networks. Int. J. Comput. Intell. Theory Pract. **5** (2010)

15. Kargl, F., et al.: Secure vehicular communication systems: implementation, performance, and research challenges. IEEE Commun. Mag. **6**, 177 (2008)

16. Al Gaith, K. (ed.): Analysis Radio Access Technology RFID/IEEE802.11p For VANET's. RFID Technology: Design Principles, Applications and Controversies, Nova Publishing, pp. 51–70 (2018)

17. Hao, X., Hortelano, J., Sakiz, F., Sen, S.: Ad hoc networks survey paper a survey of attacks and detection mechanisms on intelligent transportation systems: VANETs and IoV. Ad Hoc Netw. **61**, 33–50 (2017)

18. Issariyakul, T., Hossain, E.: Introduction to Network Simulator NS2 (2012). ISBN 9781461414063

19. Kolici, V., Oda, T., Sugihara, Y., Spaho, E., Ikeda, M., Barolli, L.: Performance evaluation of a VANET simulation system using NS-3 and SUMO considering number of vehicles and crossroad scenario. In: Proceedings of the - 2015 9th International Conference on Innovative Mobile and Internet Services in Ubiquitous Computing, IMIS 2015, pp. 22–27 (2015)

20. Behrisch, M., Bieker, L., Erdmann, J., Krajzewicz, D.: SUMO - simulation of urban mobility - an overview. In: Proceedings of the 3rd International Conference on Advances in System Simulation, pp. 63–68 (2011)

21. Sedjelmaci, H., Senouci, S.M., Ansari, N.: Intrusion detection and ejection framework against lethal attacks in UAV-aided networks: a Bayesian game-theoretic methodology. IEEE Trans. Intell. Transp. Syst. **18**, 1143–1153 (2017)

22. Mane, A.A.: Sybil attack in VANET. Int. J. Comput. Eng. Res. **06**, 60–65 (2016)

23. Yan, S., Member, S., Malaney, R., Nevat, I., Peters, G.W.: Optimal information-theoretic wireless location verification. IEEE Transic. **63**, 3410–3422 (2014)

24. Ayoob, A., et al.: Hierarchical Growing Neural Gas Network (HGNG)-Based Semi-Cooperative, 1st edn. Lambert Academic Publishing (2018). ISBN 978-613-9-89443-7

25. Ayoob, A., et al.: Performance Analysis of Routing Protocols for Mobile AD-HOC Network, 1st edn. Lambert Acadamic Publishing (2018). ISBN 978-613-9-93266-5

26. Al, G.: RFID Technology: Design Principles, Applications and Controversies. Nova Science Publishers, Inc. (2018)

Ad-Hoc Framework for Efficient Network Security for Unmanned Aerial Vehicles (UAV)

Md Samsul Haque$^{(\boxtimes)}$ ⓘ and Morshed U. Chowdhury

Deakin Centre for Cyber Security Research and Innovation,
School of Information Technology, Deakin University, Geelong, Australia
{mshaq,morshed.chowdhury}@deakin.edu.au

Abstract. With the emerging new applications, UAVs are incorporating to our daily lifestyle. The convenience of offering certain services via UAV using its cyber capabilities is very attractive but on the other side poses a great threat of safety and security. With the ever-growing use of commercial WiFi based UAVs, the enduring ability for cybersecurity and safety threats has become a sophisticated problem. UAV networks are susceptible to several common security risks including eavesdropping and jamming attack, denial of service (DoS) or buffer overflow attack by malicious remote attackers. Because of the unique nature of UAV networks, traditional security techniques used by conventional networks is not feasible for UAV communication. The resource constraint nature of such WiFi based UAV network is a key design problem, when implementation of security. Address security issues for UAV domain and propose to examine the practicality of using Identity Based Encryption (IBE) in resource constraint UAV network is our aim in this study. We would like to assess the practicality and performance of IBE in UAV network by measuring energy of the operations for key management and examine the feasibility of the approach and thus present an efficient security framework for resource constrained wireless UAV network.

Keywords: Identity-based encryption · Unmanned aerial vehicle · Omnet++

1 Introduction and Motivation

The progression of computing technology leads to the development of new generation systems called cyber physical system (CPS). UAVs are also known as cyber physical systems, as it allows cyber world for example information and intelligence to be connected to physical world using various multifunctional sensors, actuators and communication channels [1]. UAVs are ubiquitous. Military and defense operational environments are mainly used UAVs for various tactical operations. But uses of UAVs are fast growing for scientific operations, profitable, and entertaining purposes [2]. Commercially they are valuable assets to business organizations, and various government agencies, and are becoming more and more mainstream now. Regulation and law administration agencies, transporters, aerial photojournalists, humanitarian agencies and disaster management agencies, multimedia broadcasting and agriculturalists use UAV as a key instrument for their operation [2, 3]. There will be approximately

© Springer Nature Switzerland AG 2019
R. Doss et al. (Eds.): FNSS 2019, CCIS 1113, pp. 23–36, 2019.
https://doi.org/10.1007/978-3-030-34353-8_2

seven million UAVs flying United States airspace alone, as forecasted by Federal Aviation Authority (FAA) by 2020. Among these potential new UAVs, roughly three million will be flying commercially as unmanned aerial vehicles and around 4.3 million will be used for recreational, educational and scientific purpose [4]. Big companies including Amazon, Facebook, Google, UPS, FedEx, and Dominos [5] have experimented UAVs with commercial applications. In Germany, in the island of Juist, DHL successfully using UAVs to deliver medicine [6]. But the lack of developments of security frameworks and complex regulatory environments, the widespread and extensive utilization of commercial UAVs are yet to be achieved [7]. Security and privacy is the greatest concern for the 75% global members of ISACA, a poll conducted in 2017 toward the use of UAVs for business [8].

UAV networks are ad-hoc network comprised of mainly with base stations and limited resource UAVs. While on operation UAVs are controlled and monitored from the ground-controlled station, based on information sent from onboard equipment through a wireless channel. These networks are open in nature and lack of physical infrastructure. Possible interception or eavesdropping causes serious security complications due to the deployment in exposed, unattended environments [2] and hence and could be overwhelming for the safety and security of people's life and infrastructure. An example of UAV cyber-attack was deliberately jamming wireless communication link. The operator was shooting an Australian triathlon using UAV. He lost the control of the unmanned aerial vehicle while filming and it is believed to be by means of attacking the communication frequency between ground controller and the UAV. The attacker consciously interfered the operation resulting the UAV crashed onto the athletes [9]. Using simple GPS and WiFi jammers suck attacks can be performed [2]. Another attack was performed by Iraqi militants on U.S. Reapers UAV, which was used to capture real-time video feed captured by the UAV and sent to the ground control system, using an inexpensive, mass-produced product named "SkyGrabber". The attack was possible because encryption was disabled on the UAV for performance reason [10].

In 2011, Iranian cyber warfare unit claimed custody of a U.S military UAV called RQ 170 Sentinel [10]. Iranian forces used sensor spoofing attack and jammed the satellite communication and GPS functionality of UAV. The forces then easily attacked the UAV GPS system. Even though the threat was on a military UAV, but the attack can be performed on a commercial UAVs using similar technology. The very common networking services like ftp and telnet service, in the context of civilian UAV the AR. Drone, these services ports are open, and even not protected by any password. Further the unencrypted communications link allows the UAV to hijack the device, eavesdrop on video streams, and even track people [11]. All the early versions DJI Phantom UAV's it was possible to change the SSID of the access point in midair. It allows the attacker causing the UAV to disconnect from the ground controller, and allowing another device to take control of the flying machine [12].

UAVs also have several defining features such as, they have miniature sensors with omni-directional wireless communications ability, and constrained computational capability with limited battery power supply. We face many challenges with resource limitations in terms of power, computing resources, and bandwidth while the operation of wireless UAV network. For wireless communication, the encryption standard used

are called Wired Equivalent Privacy (WEP), WiFi Protected Access (WPA) and WiFi Protected Access 2 (WPA2). However, this encryption standard has vulnerabilities which can be exploited [13]. SkyJack [14] project uses a Raspberry pi, a low-cost credit card sized computer with Perl script and some free tool to intrude the UAV. To control the UAV, it then exploits the unencrypted WiFi communication channel between UAV and the smart phone. The design of frames in the IEEE 802.11 specification, made it possible for an enemy to inject de-authentication codes to the network causing the legitimate user to disconnect from the UAV network. Afterwards the attacker can gain control over the device. Because there was no encryption in place, this attack is achievable and quite easily conductible. Thus, to address a security framework is necessary for these resource-constrained UAV network over the weakest wireless links.

We thus articulate that the IBE can be effectively utilized to address such challenging scenario. Using our proposed security framework, we hope that commercial UAV manufacturers can design their security mechanism and consider the opportunities to develop their products and enhance the security against potential malicious intimidations. The rest of this paper is organized in the following sections: an overview of UAV platform and security vulnerabilities of UAV network has been provided in Sect. 2. Section 3 discusses related research in UAV technology and network. In Sect. 4 we review the platform and security issues surrounding a commercial UAV. In Sect. 4, we identify various encryption techniques to build a security network and we present IBE for key distribution for UAV network in terms of energy efficiency. We have introduced our multilayer security framework in Sect. 5. We describe our experimental evaluation in Sect. 6, and finally conclusion and directions for future work has been presented in Sect. 7.

2 UAV Platform and Security Consideration

Understanding the capabilities as well as the UAV technology platform is very important to apprehend the challenges they create and to analyze the vulnerabilities. Figure 1, shows the main components of a UAV and Fig. 2, show UAV system with information flow with environment and the ground control system (GCS).

The base system is the foundation of the UAV that links other components. The WiFi data link is also defined as WLAN (wireless LAN). The IEEE 802.11 standard based communication which ranges from 5.15 GHz to 5.75 GHz with 54 mbit/s, transmission rates used to carry control information between the UAV and the GCS. UAVs are multirotor aircraft operated by a controller. Organizations like NATO, NASA and DoD has categorized UAVs based on weight, altitude, or speed but it differs among the organizations. Common category of UAVs, based on weight and range are defined as: Nano, Micro, Mini, Small, and Tactical. These classifications based on weight, payload, altitude and range are presented in Table 1 [5].

UAVs are usually allowing video piloting capabilities based on radio signals where live video streams are sent from the UAV's video camera to the receiver at the GCS. However commercial UAVs are expected to adopt human-on-the-loop control model where, all decisions are taken by a remote human operator [16]. In this paper we focus on Micro UAV platform which has been categorized for commercial UAVs. At the

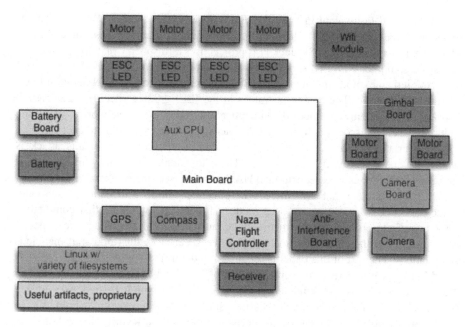

Fig. 1. Major components of an UAV systems [15].

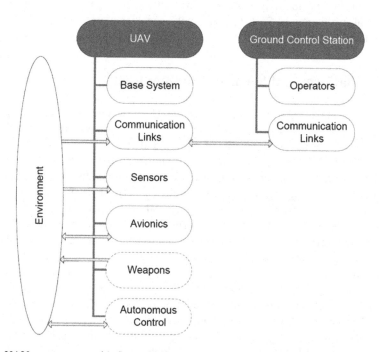

Fig. 2. UAV components with flow of information between ground control system (GCS) and the environment [10].

Table 1. UAV category based on weight, payload, range and altitude

Category	Weight	Operating altitude	Range	Payload
Nano	<0.2 kg	<90	90 m	<0.2 kg
Micro	0.25–2 kg	<90 m	5 km	0.2–0.5 kg
Mini	2–20 kg	<900 m	25 km	0.5–10 kg
Small	<150 kg	<1500 m	50–100 km	5–50 kg
Tactical	>150 kg	<3000 m	>200 km	25–200 kg

beginning UAVs were built by amateurs and used primarily by hobbyists for fun. But in 2010, a smartphone pilotable UAV, AR.drone, utilizing 802.11 wireless technology was introduced by the French company named Parrot S.A. The onboard 802.11 WiFi modules allows a client controller like a smartphone or a tablet computer or a dedicated physical controller called SkyController to bind and control the aircraft. UAVs collect and process a wide range of information that make themselves extremely interesting target for attacks, theft and data manipulation.

On the basic model of AR.Drone, has the omnidirectional wireless antenna and with usually unencrypted wireless link [10]. The Parrot's AR.Drone 2.0 runs a 1 GHz 32-bit ARM cortex A8 processor with DSP video 800 MHz and has 1 GB DDR2 RAM at 200 MHz on Linux 2.6.32 OS. It uses WiFi b/g/n for communication [17]. Since the initial development there are multiple iterations of AR.Drone software and most notably Bebop 2 airframes. The Parrot Bebop 2 is a P7 dual-core CPU and quad-core GPU embedded system running a Linux-based OS with a built in Wi-Fi a/b/n/ac, signal range up to 2 km with the Parrot SkyController 2, without obstacles or interference, GPS, and camera. It has 8 GB internal flash memory, 3350 mAh removable battery with 30 min battery life [18]. Because of the easy to use ability, light weight specifications and low cost it is very popular among the consumers. Also due to its popularity and open network communication architecture Parrot Bebop is always a good candidate for network and device security research [19].

3   Related Research

By design UAVs are relatively easy to intercept and hack. Their design considered to have fast and informal setup, messages are transferred unencrypted and many communication ports like telnet and ftp are left open as a result. Unfortunately, civilian, hobby space and commercial purpose UAV vendors mainly focusing on practical characteristics of UAV and are not aware of the underlines the security requirements. A lot of UAVs were not really designed with security in mind except for the law enforcement or military used UAVs. It is interesting that there is more research has been conducted regarding the vehicular ad hoc network for car-to-car infrastructure compare to UAV network. It is unclear if this lack of research is because of these devices are presumably considered to be secure and reliable. Some researchers examined UAV security related issues. For example the Federal Trade Commission [20] researchers were able to intercede into three different off-the-shelf UAVs namely

AR Drone Elite Quadcopter, the Hawkeye II 2nd FPV Motion Sensing Quadcopter, and the oneCase CX-10w. They were able to take over the video feed on all three of the drones, since the data was sent unencrypted. They demonstrated how they were able to connect to one of the drone's camera feeds from any computer, since the UAV's WiFi access point wasn't password protected.

The research in article [6] discussed the general overview of current attacking methods. This research focuses on defense and trust strategies to overcome cyber-attackers on UAV. WiFi based commercially available Parrot AR.Drone was used to implement the hacking procedure for this research and its severe results are depicted on this paper. Irreversible damage can be made to the UAV by conceding the communication network between the ground control system and UAV. Although this paper describes encryption, neural networks, and fuzzy logic based IDSs techniques to defend attacks but failed to discuss the details about the protocols and algorithms used. It also lacks the comparison of the performance to the solutions used to prevent the attack.

Proposing a cyber-security threat model security threats for UAV Systems are analyzed in Paper [19]. The researched model help UAV system designers, the capability to understand cyber-security threat profile of the system. This research helps to identify high priority threats, address various system vulnerabilities, and select mitigation techniques for these threats accordingly. The researchers also tried to evaluate risk by producing different vulnerabilities of the UAV system. Although several security threats to UAV is analyzed and possible attack paths has been proposed on this paper using a cyber security threat model, it is undecided which security threats might mostly distress the UAV systems.

Giary [22] proposed a hierarchical Identity Based Key Management Scheme in Tactical Mobile Ad Hoc Networks. Author offered a technique of key management in distributed hierarchal network. Hierarchical nodes can get their keys updated either from a threshold sibling or from their parents. The technique of dynamic node selection formulated as a stochastic problem and the proposed scheme can select the best nodes to be used as public key generator (PKG) from all available ones considering their security conditions and energy states. Simulation results show that the proposed scheme can decrease network compromising, probability and increase network lifetime.

Author revealed commercially available Wi-Fi-based UAVs like Parrot Bebop are vulnerable to three basic security attacks in paper [7]. These attacks are Denial of service (DoS), Buffer overflow, and ARP (Address Resolution Protocol) Cache Poisoning. Penetration test on Parrot Bebop UAV discovered that it is vulnerable to basic ARP Cache Poisoning attack, which can disconnect the user and causing the UAV force landing. Authors argue that their Multi-layer Security Framework is a direct consequence of security assessment of the Parrot Bebop UAV, and debates that the issue to be the Controller-to-UAV internetworking problem. Also, they propose that this problem can be mitigated by adding timer to limit the time the CPU can be used for non-navigational processing, hardline input data filtering and anti-spoofing mechanisms to the UAV's access point. But this research does not highlight the how this security framework will affect other commercially available UAVs.

A lot of security modelling research have been taken place for wireless sensor networks, mobile ad-hoc networks but they cannot always be applied to systems in a complex and critical infrastructure like UAV. There is a need for comprehensive

security framework for commercial UAVs, but to date none exists. So far, all the research conducted in UAV security domain, researchers bring general security awareness regarding security issue within commercial UAVs but failed to propose a complete solution. For example, in the research paper [21], the author focused on securing the access point, while the authors in paper [22] emphasized on securing the UAV's GPS. Basic security concerns for UAVs have been addressed by some researchers on article [5] whereas other researchers acknowledge that every facet of commercial UAVs lack security [11] but they stop suggesting a security framework. Therefore, it is necessary to revisit the conventional UAV security mechanism and develop an efficient security solution to fill the gap and protect future UAV wireless network.

4 Encryption Techniques

The communication in UAV network is wireless in nature. Every message transported between UAV nodes should be encrypted and authenticated. There are many methods in the literature for implementing security using cryptography, but we look briefly symmetric and asymmetric key encryption mechanism. In symmetric key cryptography, we use the same secret key for both the encrypting and decrypting information that provides data integrity and confidentiality for securing stored data. It reduces or has low computational cost and no certificate needed for secure communication. Unfortunately, it is not scalable and tends to demand high communication overhead to establish the secret key [23]. The encryption process, in asymmetric encryption uses different keys for encryption and decryption. Although the keys are different, mathematically they are related and hence decryption is possible. Asymmetric key encryption is also called public-key encryption or Public key cryptosystem (PKC), as it uses a public-key is for encryption of data. It uses the public key in public repository and the receiver uses a well-guarded secret private key to decrypt the message. These keys are mathematically related but computationally not feasible to find one from another and thus PKC provides the most effective mechanisms for establishing security. This scheme of encryption is scalable and provide authentication, non-repudiation, integrity, confidentiality of data with undeniably computationally intensive large key size [24].

4.1 Applying IBE to UAV Network

One of the major challenges with symmetric key cryptography is to establish a way to share the key to the other party for message transfer. Symmetric encryption keys are often encrypted with an asymmetric algorithm like RSA and sent separately. Identity-Based Encryption (IBE) can solve this problem. With IBE, there is no need to generate a public/private key pair. The public key is not circulated, instead this key can be an arbitrary identity string such as IP address, an email address or any other identity info. Thus, eliminates the need for a certificate authority to provide an assurance of the holder between a public key and of the corresponding private key. And decreases the frequency of key negotiation and reduces the communication overheads for resource constrained UAV network. Shamir [25] proposed the idea of the IBE scheme which is

based on public-key cryptography and uses unique ID of the device as its public key. Boneh and Franklin [26], in 2001 described the first practical IBE system. The most efficient identity-based encryption scheme currently based on bilinear pairing on elliptic curves. Basic Boneh-Franklin IBE scheme consists of four randomized algorithms [27] namely setup, extract, encryption and decryption. These four subsystems will assume that a randomized algorithm $\mathcal{G}$ takes a security parameter $k \in Z$ outputs a prime number q, plaintext space $M = \{0, 1\}^n$ and ciphertext space is $C = G_1 \times \{0, 1\}^n$. The following section describes the basic Boneh-Franklin IBE implementation:

(1) Setup: In this sub system the master-key is generated, and system parameters are set. The description of the message space and the cipher text space are included in the parameters. Except the master key which will be kept secret and is only known to the PKG all other system parameters will be publicly known. The following three steps will be performed for setup sub system.
(a) Run $\mathcal{G}$ on input k to generate a prime q, two groups G_1 and G_2 with order q, random generator $P \in G_1$, as well as a bilinear mapping $\hat{e}: G_1 \times G_2 \rightarrow G_2$.
(b) Pick a random master key $s \in Z_q^*$ and compute $P_{pub} = sP$.
(c) Select two hash function $H_1 : \{0, 1\}^* \rightarrow G_1^*$, $H_2: G_2 \rightarrow \{0, 1\}^n$ for some n. The security analysis will view H_1, and H_2 as random oracles. The system parameter $\pi = \{q, G_1, G_2, \hat{e}, n, P, P_{pub}, H_1, H_2\}$, master key is $s \in Z_q^*$.
(2) Extract: Generation of the private key specific to every node corresponding to a public key identity string and the master private key of the system. For a given string $ID \in \{0, 1\}*$, compute $Q_{Id} = H_1(ID) \in G_1^*$, and compute $K_{Id} = (Q_{Id})^s$ as private key for corresponding user where s is the master key.
(3) Encryption: Producing the ciphertext of a plaintext message using the public master key and the public key identity string. For a plaintext $m \in M$, and its ID, the method to encrypt it is to: first, compute $Q_{Id} = H_1(ID) \in G_1^*$; second, choose a random number $r \in Z_q^*$; third, construct ciphered text.

$$C = <rP, m \oplus H_2(g_{ID}^r)> , \text{among which } g_{ID} = \hat{e}(Q_{ID}, P_{pub}) \in G_2^*$$

(4) Decryption: Producing the plaintext message of a ciphertext using the corresponding private key, of the sensor node and the public master key. If the ciphered text $C = <U, V>$, encrypted using the public key ID. To decrypt C using the private key $K_{Id} \in G_1^*$ to compute the plaintext $m = V \oplus H_2(\hat{e}(K_{ID}, U))$.

5 Proposed Security Framework

Adhoc networks can be classified as flat and hierarchical topology structures. Hierarchical structures split the network into several clusters. These arrangements enhance the network quality of service, strengthen network scalability and maximize network

throughput simultaneously than flat structures with hundreds of nodes in the network. Our proposed model provides lightweight construction and security by proposing a hierarchical structure that off-loads computationally expensive workloads from resource constrained devices to powerful equipment. To achieve the goal, we want to leverage the flat architecture of the underlying system and construct a multilevel hierarchical structure.

5.1 Hierarchical UAV Network Architecture

In UAVs we use wireless networks for data communication to facilitate the interaction between base station and the UAV. These networks are exposed and unguarded. So, potential interception or eavesdropping can cause security concerns and it is possible for an adversary to snoop or fabricate the transmitted information. Also, these sensors are restricted in terms of bandwidth, energy, computing power, storage, and memory. These resources constrained nature make it impractical for WSNs to deploy traditional security schemes to transmit data between UAVs. Therefore, we are proposing a multilevel hierarchical system for data transfer that distributes computing overhead in UAV ad-hoc network and at the same time ensures the independence and security of the sub-networks. The approach in hierarchical system is different than a classical flat network system in which all cluster members have the same access rights. In hierarchical network systems access to information brings a new level of security to the system. Information can only be viewed by those who have access to it. A hierarchical structure of UAV network for load balancing and overhead distribution is shown in Fig. 3.

A network is divided into cluster or groups in a hierarchical structure. Each cluster has a cluster head and connected to upper or lower layers directly or indirectly. It is possible for each cluster to assign different functionalities or operate in specific areas, with a regularly and dynamically nominated cluster head (CH) for each cluster. The ordinary cluster member (OCM) which is a member in a specific cluster, communicate only with its CH. The OCMs or nodes has different attributes such as energy usages capabilities, size etc. The base station (BS) chooses N nodes to be CH, which forms N sub-networks during network initialization. The CH chosen by BS delimit their territories and broadcast their identities to OCMs within their boundaries. The CHs transmit information received from OCMs to BS and make a fundamental information fusion. If the base station is accessible to CHs, cluster heads transmit messages directly to BS, if not, CHs make a multi-hop transmission via reply nodes. The CH is superior to OCMs. They have more computational ability, storage, memory, and energy and battery power. Cluster heads performs tasks such as aggregating information from the OCMs, processing data within the cluster, forwarding the data to base station and leading the cluster to the destination. For new node addition, it registers its identity to the CH, where it belongs to. CH checks the validity and authenticity of the node from BS. If a node needs to be revoked or compromised, the CH discards the corresponding communication and reports the negated node to BS.

To our knowledge, no other related works propose a multilayer approach to securing Flying ad hoc UAV networks using IBE encryption mechanism. For example Flying Ad-Hoc Networks (FANETs) are surveyed in [28] as an ad-hoc network

BASE STATION

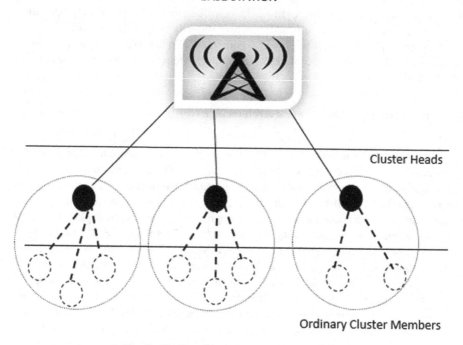

Fig. 3. UAV network clustering architecture.

connecting the UAVs. The author introduced FANET simulation testbeds by discussing various ad-hoc network and the main design challenges of FANET. But this research does not discuss about FANET security and resource constrained nature of UAVs. Research study in paper [29], described a security framework for UAV assisted multilevel ad-hoc wireless networks, based on identity-based public key infrastructure using bilinear pairing, consists of group key management architecture, static pair-wise communication, tripartite key agreement, and group key agreement. Their proposed design reduces the certificate validation process, improves computational efficiency and reduces storage requirement. This security framework provides an efficient and flexible security service suitable for mobile ad-hoc network. Although this framework can easily and quickly adapt to other ad-hoc network, but it does not provide any evidence or evaluate the performance of IBE uses. As like the previous research, authors of [30] proposes a distributed hierarchical key management scheme using IBE in which nodes can get their keys updated either from their parent nodes or a threshold of sibling nodes and simulation results show that the proposed scheme can decrease network compromising probability and increase network lifetime in tactical MANETs. This research also fails to adhere the characteristics of ad-hoc UAV network and does not provide the evidence resource consumption.

6 Experimental Evaluation

We have presented a survey and various security issues in UAV wireless network so far in this paper. Our contribution to this area of network security is to analyze the feasibility of using IBE for resource constrained UAV network. We would like to examine the power and energy requirements for IBE to make it feasible for UAV network. To complete our experiment, we need a software to simulate UAV wireless network, implement identity based asymmetric key encryption system, and coding to measure efficiency by calculating power consumption of IBE in UAV network communication.

We have used an Intel Core i7-3770 processor CPU, with a high-end NVIDIA GTX 580 GPU on a host machine as the setup for the experiment. Using virtual box, we have built an Ubuntu 18.0402 LTS virtual machine. To conduct detailed experiment, we are using an off the shelf simulation software omnet++. OMNET++ simulation uses C++ class library. Messages in omnet++ represent frames or packets in a computer network [31]. We will be using INET Framework model library for the omnet++ simulation environment to calculate power consumption of IBE. In INET, an energy consumer model is an omnet++ simple module that implements the energy consumption of software processes to provide the power or current consumption for the current simulation time. Our target UAV platform is Parrot Bebop with battery capacity 3350Ah and flight time up to 30 min. The following two figures, Figs. 4 and 5 shows a basic UAV network simulation in omnet++ and NED source file for the experimental setup. We will further improve this design by adding more nodes, channel modules and power modules. Using C++ programming will create.ini files for communication between UAVs and IBE key encryption system.

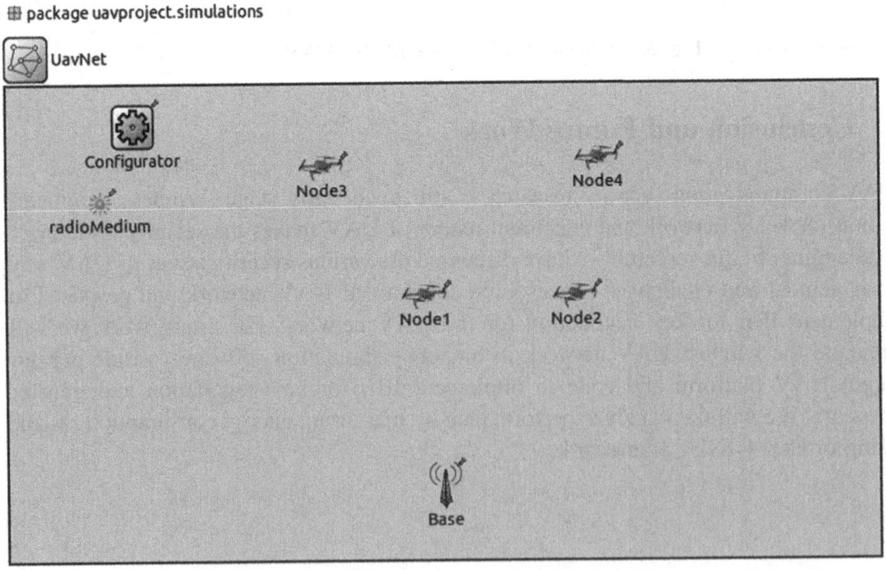

Fig. 4. UAV network simulation in omnet++.

```
package uavproject.simulations;

import inet.networklayer.configurator.ipv4.Ipv4NetworkConfigurator;
import inet.networklayer.configurator.ipv4.Ipv4NodeConfigurator;
import inet.node.ethernet.EtherHost;
import inet.node.inet.AdhocHost;
import inet.node.inet.WirelessHost;
import inet.physicallayer.ieee80211.packetlevel.Ieee80211ScalarRadioMedium;


network UavNet
{
    @display("bgb=772,408");
    submodules:
        Node1: AdhocHost {
            @display("p=371,182;i=misc/drone");
        }
        Node2: AdhocHost {
            @display("p=504,182;i=misc/drone");
        }
        radioMedium: Ieee80211ScalarRadioMedium {
            parameters:
                @display("p=83,101;is=s");
        }
        Node3: AdhocHost {
            @display("p=281,71;i=misc/drone");
        }
        Node4: AdhocHost {
            @display("p=524,64;i=misc/drone");
        }
        Configurator: Ipv4NetworkConfigurator {
            @display("p=113,35");
        }
        Base: WirelessHost {
            @display("p=377,350;i=device/antennatower");
        }
}
```

Fig. 5. Omnet++ NED source file for UAV network.

7   Conclusion and Future Work

UAV communication security research is still in its early stage. Wireless communication for UAV network and constraint nature of UAV makes the security issue more
challenging. In this research we have discussed the various security issues of UAV, key
management and challenges for resource constrained UAV network and proposed to
implement IBE for key negotiation for the UAV network. For future work we will
simulate the wireless UAV network in omnet++ simulation software considering our
target UAV platform and code to implement IBE for key negotiation and transfer
message. We will then analyze performance by measuring energy consumption of IBE
using omnet++ INET framework.

References

1. Dini, G., Tiloca, M.: A simulation tool for evaluating attack impact in cyber physical systems. In: Hodicky, J. (ed.) MESAS 2014. LNCS, vol. 8906, pp. 77–94. Springer, Cham (2014). https://doi.org/10.1007/978-3-319-13823-7_8
2. Rani, C., Modares, H., Sriram, R., Mikulski, D., Lewis, F.L.: Security of unmanned aerial vehicle systems against cyber-physical attacks. J. Defense Model. Simul.: Appl. Methodol. Technol. 13(3), 331–342 (2015)
3. Snell, B.: McAfee Labs 2017 threats predictions: "Dronejacking" places threats in the sky, November 2016, 2017. https://www.mcafee.com/au/resources/reports/rp-threats-predictions-2017.pdf
4. FAA Releases 2016 to 2036 Aerospace Forecast (2016). https://www.faa.gov/news/updates/?newsId=85227
5. Nassi, B., Shabtai, A., Masuoka, R., Elovici, Y.: SoK-security and privacy in the age of drones: threats, challenges, solution mechanisms, and scientific gaps. arXiv preprint arXiv: 1903.05155 (2019)
6. Moormann, D.: DHL parcelcopter research flight campaign 2014 for emergency delivery of medication (2015)
7. Hooper, M., et al.: Securing commercial WiFi-based UAVs from common security attacks, pp. 1213–1218. IEEE (2016)
8. Securitymagazine: Privacy and security are biggest concerns about the business use of drones, 2 March 2017. https://www.securitymagazine.com/articles/87868-privacy-and-security-are-biggest-concerns-about-the-business-use-of-drones
9. Gallagher, S.: Triathlete injured by "hacked" camera drone (2014). https://arstechnica.com/security/2014/04/triathlete-injured-by-hacked-camera-drone/. Accessed June 2017
10. Hartmann, K., Steup, C.: The vulnerability of UAVs to cyber attacks-an approach to the risk assessment, pp. 1–23. IEEE (2013)
11. Samland, F., Fruth, J., Hildebrandt, M., Hoppe, T., Dittmann, J.: AR. drone: security threat analysis and exemplary attack to track persons. In: International Society for Optics and Photonics, p. 83010G (2012)
12. Trujano, F., Chan, B., Beams, G., Rivera, R.: Security analysis of DJI phantom 3 standard. Massachusetts Institute of Technology (2016)
13. Reddy, S.V., Ramani, K.S., Rijutha, K., Ali, S.M., Reddy, C.P.: Wireless hacking-a WiFi hack by cracking WEP, pp. V1-189–V1-193. IEEE (2010)
14. Kamkar, S.: SkyJack (2013). Accessed June 2019
15. Kovar, D.: UAVs, IoT, and Cybersecurity (2016)
16. Altawy, R., Youssef, A.M.: Security, privacy, and safety aspects of civilian drones: a survey. ACM Trans. Cyber-Phys. Syst. 1(2), 7 (2017)
17. Parrot Ar.Drone 2.0 Power Edition, Technical Specifications. https://www.parrot.com/global/drones/parrot-ardrone-20-power-edition
18. PARROT Bebop 2 Power - Pack FPV, Technical Specifications. https://www.parrot.com/global/drones/parrot-bebop-2-power-pack-fpv
19. Krajník, T., Vonásek, V., Fišer, D., Faigl, J.: AR-drone as a platform for robotic research and education. In: Obdržálek, D., Gottscheber, A. (eds.) EUROBOT 2011. CCIS, vol. 161, pp. 172–186. Springer, Heidelberg (2011). https://doi.org/10.1007/978-3-642-21975-7_16
20. Valente, J., Cardenas, A.A.: Understanding security threats in consumer drones through the lens of the discovery quadcopter family, pp. 31–36. ACM (2017)

21. Pleban, J.-S., Band, R., Creutzburg, R.: Hacking and securing the AR. drone 2.0 quadcopter: investigations for improving the security of a toy. In: International Society for Optics and Photonics, p. 90300L (2014)

22. Giray, S.M.: Anatomy of unmanned aerial vehicle hijacking with signal spoofing, pp. 795–800. IEEE (2013)

23. Fang, Y., Zhu, X., Zhang, Y.: Securing resource-constrained wireless ad hoc networks. IEEE Wirel. Commun. **16**(2), 24–30 (2009)

24. Doyle, B., Bell, S., Smeaton, A.F., McCusker, K., O'Connor, N.E.: Security considerations and key negotiation techniques for power constrained sensor networks. Comput. J. **49**(4), 443–453 (2006)

25. Shamir, A.: Identity-based cryptosystems and signature schemes. In: Blakley, G.R., Chaum, D. (eds.) CRYPTO 1984. LNCS, vol. 196, pp. 47–53. Springer, Heidelberg (1985). https://doi.org/10.1007/3-540-39568-7_5

26. Boneh, D., Franklin, M.: Identity-based encryption from the weil pairing. In: Kilian, J. (ed.) CRYPTO 2001. LNCS, vol. 2139, pp. 213–229. Springer, Heidelberg (2001). https://doi.org/10.1007/3-540-44647-8_13

27. Kodali, R.K., Chougule, S.K.: Hierarchical key agreement protocol for wireless sensor networks. Int. J. Recent Trends Eng. Technol. **9**(1), 25 (2013)

28. Bekmezci, I., Sahingoz, O.K., Temel, Ş.: Flying ad-hoc networks (FANETs): a survey. Ad Hoc Netw. **11**(3), 1254–1270 (2013)

29. Chien, H.-Y., Lin, R.-Y.: Identity-based key agreement protocol for mobile ad-hoc networks using bilinear pairing, pp. 8–pp. IEEE (2006)

30. Yu, F.R., Tang, H., Mason, P.C., Wang, F.: A hierarchical identity based key management scheme in tactical mobile ad hoc networks. IEEE Trans. Netw. Serv. Manag. **7**(4), 258–267 (2010)

31. Sliwa, B., Ide, C., Wietfeld, C.: An OMNeT++ based framework for mobility-aware routing in mobile robotic networks. arXiv preprint arXiv:1609.05351 (2016)

Performance Evaluation of LoRaWAN for Mission-Critical IoT Networks

Ansa Iftikhar Ahmad[1]([✉]), Biplob Ray[2], and Morshed Chowdhury[1]

[1] Deakin University Centre for Cyber Security Research and Innovation,
Deakin University-Geelong, Geelong, VIC, Australia
ansa.ahmad@deakin.edu.au
[2] School of Engineering and Technology, Centre for Intelligent Systems (CIS),
Central Queensland University, Rockhampton, Australia

Abstract. With the evolution of wireless communication in Internet of Things (IoT) networks, Low Power Wide Area Network (LPWAN) has attracted a lot of attention due to its low cost and low power usages. Some of the LPWAN offerings are mainly proprietary but Long-Range Wide Area Network (LoRaWAN) is an open standard communication protocol (ALOHA-based) for a network using the Long Range (LoRa) in the physical layer. Due to its bi-directional communication and Adaptive Data Rate (ADR) capability, the LoRaWAN gateways are adopted in various IoT networks, like smart city, smart farming, worldwide. However, for wider adoption of LoRaWAN in mission-critical applications, it must be tested for scalability and reliability in various practical scenarios to reduce adverse impact in the system. This paper has conducted an evaluation of scalability and reliability of LoRaWAN using three practical scenarios of IoT systems. The evaluation has considered throughput performance, spreading factor statistics, gateway coverage assessment, and success probability performance of the protocol to reveal the performance of the protocol. The evaluation result shows that LoRaWAN networks are decidedly scalable supporting hundreds or thousands of end devices; however, on the other hand, there is an impression where scalability could be inversely proportional to performance only with an increased number of nodes and not gateways, thus requires a solution at the nodes. Our evaluated result can be very useful not only for designing the LoRaWAN based IoT network but also for improving LoRaWAN data transmission techniques for more reliable data transfer between sensor nodes and gateway.

Keywords: LoRaWAN · Reliability · Scalability

1 Introduction

The Internet of Things (IoT) is thriving in today's world and its usage is wide spread having the capacity to progress our everyday lives. Due to improved wireless communication techniques, wireless technologies like sensors, actuators and sensory tags are embedded into day to day and mission-critical objects like vehicles, buildings and appliances to make IoT concept a reality [1, 2]. Among the various multiplicity of wireless connectivity technologies for IoT, the Low Power Wide Area Network

© Springer Nature Switzerland AG 2019
R. Doss et al. (Eds.): FNSS 2019, CCIS 1113, pp. 37–51, 2019.
https://doi.org/10.1007/978-3-030-34353-8_3

(LPWAN) is gaining huge attention due to its unique features like wide area coverage capability, low cost, and low power features [3].

Amongst various LPWAN technologies, Long Range (LoRa) driven LPWAN (Low Power Wide Area Network) named Long-Range Wide Area Network (LoRaWAN) is widely utilized by many IoT networks which is evident from immense deployment of IoT network based on LoRaWAN around the world [3, 4]. This is since the LoRaWAN based technologies offer features like bidirectional communication, long range and low power which is useful features for both suburban and urban areas to build widespread and extensive IoT networks [5]. In LoRaWAN, the LoRa modulation in physical layer (PHY) ensures longer range capability than other legacy wireless technologies [5]. The LoRaWAN technologies offer low bandwidth, long range, energy and cost-efficient connectivity resolution for tracking devices, smart meters, infrastructure and environmental sensors [6] to use in IoT systems like Machine to Machine (M2M) communication, smart agriculture (precision cropping, smart irrigation) and smart city (on street lightening, smart metering) solutions to address power and budget issues. The LoRaWAN offers communication capability of more than 10 km between gateway and end devices with battery lifetime of 5 to 10 years [5, 7]. The trade-off is less than 5 Kbps of data rate. It has a simple network topology i.e. star which helps in cutting the network maintenance and deployment cost [7]. The LoRaWAN technology can be very useful for less data intensive long-range communication systems of mission-critical applications like smart grid, rural smart health and traffic systems. However, the mission-critical applications require reliability and time sensitivity of data transmission. Hence, it is critical to know the level of reliability and timeless of LoRaWAN based IoT systems in various practical scenarios for scalability. Therefore, in this paper, we have presented evaluation of three IoT network scenarios to present a comparison of resource usages, throughput, packet error ratio and fairness of LoRaWAN based IoT network. The result of this evaluation can be very useful for IoT network engineers to design efficient and improved LoRaWAN networks. The scientists would be able to use the result for further research on LoRaWAN communication technique for improved scalability and reliability.

The rest of the paper is organized as follows. Section 2 presents a literature review and working mechanism of LoRaWAN, which is followed by simulation setup and IoT scenarios used in this paper in Sect. 3. In Sect. 4, we present simulation results which is followed by analysis of the result and conclusion in Sects. 5 and 6 respectively.

2   Background and Related Work

In this section, we first present literature related to LoRaWAN which is followed by a detailed working technique of LoRaWAN.

2.1   Related Work

The LoRaWAN is comparatively a new protocol which is still under comprehensive development, and standardization is yet under active work in progress [8], therefore,

less work exists in the literature regarding the performance of LoRaWAN based IoT networks.

A number of papers worked on scalability of LoRaWAN networks with a conclusion that it is not easy to scale LoRaWAN networks [9–11]. Georgiou et al. [10] and Haxhibeqiri et al. [12] studied uplink traffic in ALOHA-based LoRaWAN network and stated that an increase in number of nodes results decline of packet delivery ratio promptly. While in papers [9] and [12] authors have considered only single spreading factor whereas in [11] authors mentioned that dissimilar spreading factors affect the performance most. According to opinion of authors in [13], the LoRaWAN does not scale, and higher spreading factors effects the network performance negatively. The experiment in [13] attempted to improve LoRaWAN's capability by shifting nodes to higher spreading factors which has resulted upsurges the nodes chances of collision.

The authors in [14] have studied downlink traffic of LoRaWAN and have found that downlink message transmission could corrupt the packets of uplink at the gateway. Hence, the downlink traffic should be directed with care for example the acknowledgements. The authors in [14] propose RS-LoRa to improve scalability of LoRaWAN's by a novel light weight scheduling for the gateway and a self-scheduling approach for the nodes.

The paper [15] attempted to offer more reliability to the nodes that are far away from the gateway. It is presented in this paper that uplike in LoRaWAN choose their preferred spreading factor whereas network performance could be improved expressively if the gateway assigns the spreading factors. Depending on the offered derivations, the perfect percentage of traffic utilizing (TU) of one spreading factor can be expressed as presented in Eq. (1).

$$TU_i = \frac{(\text{bit rate})_i}{\sum_{s=n}^{s=1}(\text{bit rate})_s} \tag{1}$$

Although, when a node is allocated the highest spreading factor in this scheme, so the power consumption is also high.

Authors in [16] suggested synchronizing the network LoRa completely, because ALOHA's access of the channel is the major reason for LoRaWAN's low performance. The paper [16] argued, theoretically, implementing fine-grained scheduling where a slot is assigned to each node would improve the network. However, assigning a slot to each node is very challenging for large networks such as LPWAN's as it would violate the restrictions on duty cycle.

A simple MAC protocol solution named Low-Power Distributed Queuing (LPDQ) solution is proposed in [17] and [18] where nodes in contention slots send small packets to request channel access. Nevertheless, it is impractical to directly put LPDQ on LoRaWAN, because to transmit the data packets which are significantly longer, a slot is granted to a node when the request at the gateway arrive without any collisions. However, the LoRa packets greatly vary in size, and are naturally small, from limited to 1 ms to 2 s. Therefore, outlining any slot limitations makes the scheduling complex or even waste a great number of resources that are waiting idle.

Weightless-SIG [19] and Ingenu [20] are other two approaches attempted to improve LoRa's communication. The Ingenu [20] divides the time into two slots, one for uplink and one for downlink in the 2.4-GHz ISM band. Ingenu approach is not possible in the 868-MHz band because of significantly smaller duty cycle, as the nodes in this approach want to communicate data with a random phase using a random subslot. The Weightless protocol [19] on the other hand does it's work in the 868-MHz ISM band by exploiting functionalities of TDMA/FDMA in which the gateway first sends a beacon including the nodes' scheduling. Then in the scheduled slots, the nodes would transmit their data. Weightless protocol could use a duty cycle of 67% or up to 2400 s of downlink per hour, by full usage of 868-MHz ISM band. To fulfil long range communications needs it is more than enough. Though in LoRaWAN due to wider bandwidths, it results in lower duty cycle and less potential channels over the entire band.

As detailed above, most of the work in literature review worked on improving scalability and performance of LoRaWAN network which can be benefitted by understanding the specific capability of LoRaWAN in different IoT network scenarios. Hence, in this paper, we have evaluated specific capability of LoRaWAN based IoT network.

2.2   How LoRaWAN Works

The network protocol standtard of LoRaWAN was initially developed in France by Cyclos and attained by Semtech Corporation which formed LorRa alliance and utilizes unlicensed ISM spectrum below 1 GHz. The first version of LoRaWAN standard was initially released in 2015, and the current version 1.1, in 2016. The LoRa modulation that provides longer range than other legacy wireless technologies define the physical layer (PHY) of LoRaWAN standard, and LoRaWAN is data link layer (MAC) protocol which defines the system architecture and communication protocol. The LoRa uses Chirp Spread Spectrum (CSS) modulation techniques. The MAC mechanism in LoRaWAN standard, utilizes the LoRa modulation to allow communication of many end devices with a gateway.

A Possible 6LoRaWAN network architecture is depicted in Fig. 1. A typical network architecture of LoRaWAN is systematized in a start of stars topology as illustrated in Fig. 1. The end devices communicate with the gateway, over a single wireless hop utilizing LoRa with LoRaWAN. LoRaWAN frames are forwarded to a central network server by the gateway from the end devices through a higher throughput non LoRaWAN network, typically Wi-Fi, satellite, Ethernet or 3G/4G. The gateway handles traffic bi-directionally. The network server that receives packets from the end-devices is accountable to decode them and send information to action in the application. The LoRaWAN terminology is composed as follows [21]:

- **End device/node** – an object which is implanted with low power communication device.
- **Gateway** – antennas that send data to end devices and receive broadcasts from end devices.
- **Network server** – server that directs messages from end devices to the designated application, and back to the end device from the application.

- **Application** – it is a software, that is running on a server.
- **Uplink message** - message transmitted from the device to an application
- **Downlink message** – message transmitted from an application to the device.

LoRaWAN defines three communication functionalities, and any LoRaWAN devices utilized in the network must be configured into one the three classes [22].

- **Class A**: it is the default class that allows bi-directional communication. The devices in this class are allocated a set time for communication. This class consumes the least amount of energy. It is always that end devices start the communication with the server.
- **Class B**: in addition to an allotted time to send messages, devices in class B are also scheduled to receive messages from the gateway.
- **Class C**: nodes or devices in class C continuously receive windows since it keeps on listening for messages from the gateway, it consumes the most energy.

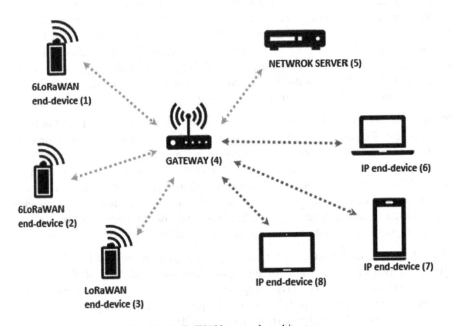

Fig. 1. LoRaWAN network architecture

An existing LoRaWAN network could be elevated with 6LoRaWAN devices i.e. IPv6 devices, if the network server or the gateway supports the communication with 6LoRaWAN and LoRaWAN protocol in the same network over same devices. As illustrated in Fig. 1, the end devices 1 and 2 represent 6LoRaWAN and a single end device number 3 represents a native LoRaWAN. They all communicate with the same number 4 gateway. The gateway that transmits all packets to the network server number 5 and could operate as a pure packet forwarder. Moreover, the gateway also handles the 6LoRaWAN packets and transmit this to IP end devices number 6 to number 8 over an IP network.

3   Simulation Setup and Scenario

This section covers the method, design and the architecture of the LoRaWAN network in ns-3 in Subsect. 3.1 which is followed by IoT network scenarios in Subsect. 3.2 used to conduct evaluation.

3.1   Simulation

In ns-3 [23], we have implemented LoRaWAN standard using modular concept in two layers named: physical and MAC layer which are ns-3 LoRaWAN modules that are added to simulate the transmissions. The discussion starts from physical layer, next MAC layer and lastly use of these modules in ns-3 (the application layer). In general, the physical and the MAC layer implementations are connected to the ns-3 call backs standard to communicate with the rest of the network stack. It allows to use a substantial amount of features within the framework of ns-3. To automatically connect all callback functions required, a helper class is provided which allows the LoRaWAN NetDevice to be configured easily with a default set of parameters. However, these parameters could also be altered utilizing the corresponding interfaces [24]. This LoRaWAN module simulates the LoRa network behavior with their applications from the devices, the features and its channel, such as shadowing and losses. Specific detail of physical and MAC layers further detailed below.

Physical Layer. The physical layer is consisted of two classes: LoRaGwPhy a physical layer for gateway and LoRaPhy a physical layer for node. LoRaPhy on every single node, keeps track of all the noise occurrence on every channel and the spreading factors used. The implementation for LoRaGwPhy is slightly different from the physical layer for nodes, where the gateway simultaneously receives multiple packets. A derived class of LoRaPhy is implemented for LoRaGwPhy. Both classes could be differentiated regarding receiving part, because LoRaGwPhy keeps a track list of all LoRa signals incoming rather than only adding those to the noise.

MAC Layer. Like Physical Layer the implementation of the MAC Layer is also divided in two classes because of the uplink and downlink of LoRaWAN. LoRaNetDevice consists MAC layer implementation of the node, whereas, LoRaGwNetDevice implements MAC layer for the gateway. Also, like physical layer the gateway class of MAC is also derived from the node class.

After implementation of these classes in the network model, the structure process followed in ns-3 to create the scenarios for evaluation:

1. **Created the topology** – To create required IoT network, collections of node objects end devices and gateways are created in the desired number. The network server is created to establish a node's position in the simulated space.
2. **Built the Model** – After creating specific number of end devices and gateways, the LoRaWAN protocol stack is installed on every node.
3. **Configuration of the network** – we configured the LoRaWAN protocol standard to use values specific for a simulation, for example whether to use ADR, the maximum number of transmissions allowed, and the message type. The LoRa

channel with the presence of buildings and shadowing's was created using propagation loss desired. Furthermore, the point to point links to connect the gateways and network server was created.

4. **Execution** – The simulations are executed after the code has been complied and run in the ns-3 software. The simulation first executes events and the conforming function calls. While the simulations were running, trace source the data in relevant data structures.

5. **Performance analysis** – The data saved by trace sources is finally visualized and analyzed.

The next sub-section detailed three scenarios used for this evaluation.

3.2   Scenarios

The paper has used three scenarios to cover wide range of the possible real life IoT network design using LoRaWAN. All these scenarios vary in their transmission techniques and gateway numbers used in the network. All three scenarios were simulated three times using 100, 500 and 100 end nodes. These three scenarios are detailed below.

- Scenario one– In the first scenario we have used one gateway where all the nodes send data to a central gateway. As illustrated in Fig. 2, the end nodes are spread in a circle around the gateway. and send unconfirmed data to the network server. The sender end nodes do not wait for any acknowledgements. However, the gateway responds within every 96 messages to avoid the nodes switching higher spreading factors.

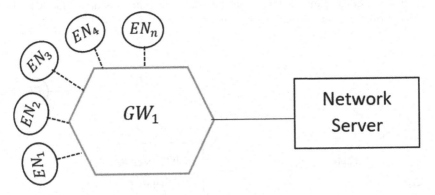

Fig. 2. Scenario one of LoRaWAN IoT network

- Scenario two – As illustrated in Fig. 3, in the second scenario instead of a central gateway; we have used multiple gateways, seven to be specific. This scenario is used to depict space diversity of LoRaWAN where different gateways could connect to the network and several gateways could respond to different nodes in the network. Like first scenario, in this scenario, the end nodes do not wait for acknowledgements.

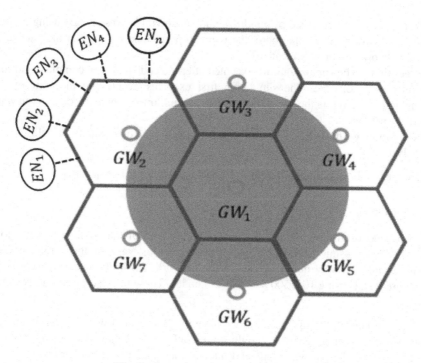

Fig. 3. Scenario two of LoRaWAN IoT

- Scenario three – In the last scenario depicted in Fig. 4, the IoT network uses multiple gateways and all end nodes transmit confirmed messages to depict a network where every message is acknowledged. The end nodes and gateways depict the behaviors of scenario 1 and scenario 2 respectively.

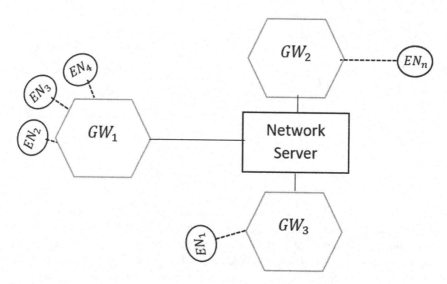

Fig. 4. Scenario three of LoRaWAN IoT

4   Results

This section presents the result of our simulation on performance of the IoT network for all three scenarios detailed in Sect. 3.

4.1   Single Gateway (Scenario One)

In the scenario one, which is single cell scenario, all nodes had to contend to transmit their data to the only gateway GW_1 due to the ALOHA nature of the protocol. However, our research is established within the delimits of pure MAC AOHA protocol, and in future we would be performing a revival research by substituting Slotted ALOHA (S-ALOHA) where the results could perform the pure ALOHA deployment results by possibly attaining a substantial increase I network downlink bandwidth, regardless of performing no changes to the standard LoRaWAN firmware and at no extra cost in hardware. The simulation result of scenario one illustrated in graph in Figs. 5 and 6. The Fig. 5 represents the reliability as a function of the distance where x-axis represents distance of the end nodes from gateway and y-axis repsents packet error ratio in percentage(%). It is clear from the illustration of Fig. 5, the reliability is increased packet error ratio is decreased when the distance to the gateway is less as the receiving power is high. This result is due to the capture effect, as it lets the packets with high power survive collisions, where as the packets with low power are discarded. As expected, more devices results in a higher packet error ratio.

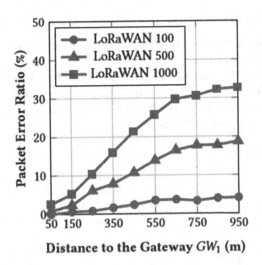

Fig. 5. Reliability vs distance in scenario one

The Fig. 6 depicts the total number of packets that are transmitted by all the nodes, and the number of packets that are successfully received by the GW_1 in scenario one. The dark blue bar at the left shows all transmitted messages counted. We have run the simulation for 24 h as illustrated in Fig. 6 where x-axis represents number of nodes and

y-axis represents transmitted and received packets in 24 h. The light blue bar at the right shows the successfully decoded packets at the gateway, hence, the amount of packets loss is the difference between the two bars. The degradation of the network is due to the random access of the share channel in the network.

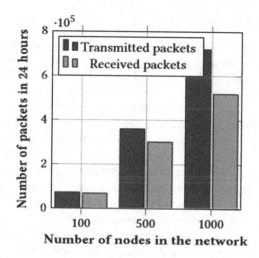

Fig. 6. Packets loss in scenario one

4.2   Multiple Gateways (Scenario Two)

In second scenario, the IoT network topology uses seven gateways. The simulation result of second scenario is presented in Figs. 7 and 8.

The Fig. 7 depicts 30% lower packet error ratio than Fig. 5 due to use of multiple gateways. Furthermore, the packet error ratio is very low where distance to the gateway is very less, and vice versa. Nevertheless, after 500 m, nodes benefit from the capture effect because they get closer to the network.

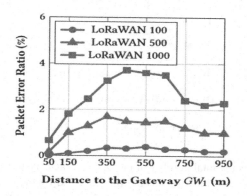

Fig. 7. Reliability vs distance in scenario two

The Fig. 8 also depicts that the reliability is higher in the multiple gateway scenario, because when a collision occurs between two nodes, it is very unlikely that both of them are at the same distance from the gateway; so one of the packets sent from the nodes will be received by the central gateway but the other packet would be discarded; but when there are multiple gateways in the network topology, the packet that is discarded would be received by a neighboring gateway and would be successfully decoded, therefore the network performance is improved.

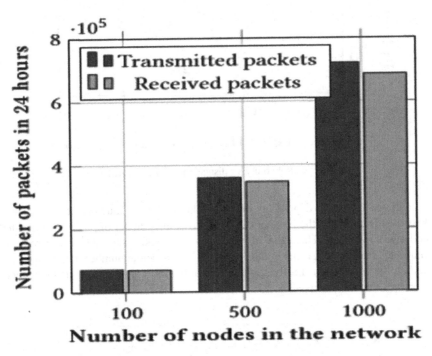

Fig. 8. Packets loss in scenario two

4.3   Reliable Transmission with Multiple Gateways (Scenario Three)

In the third IoT network scenario, multiple gateways were connected where for each transmission acknowledgements are transmitted. The simulation results of scenario three presented in Figs. 9 and 10.

The graph in Fig. 9 shows low packet loss for a small number of devices. The packet error ratio is 0 in all cases, for a network with 100 nodes and the reliability drops when the number of nodes increases. The drastic increase in the packet error ratio, deteriorating the channel far away from gateway is explained in Fig. 10.

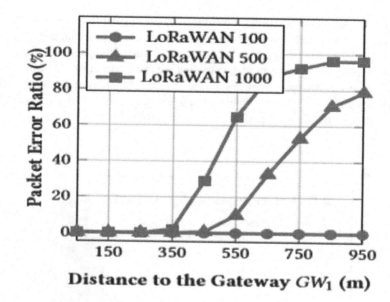

Fig. 9. Reliability vs distance in scenario three

The graph in Fig. 10 depicts the low reliability due to so much acknowledgement packets overhead. When the number of nodes grow, the number of packets transmitted by nodes also increases. This value of the transmitted packets should be compared with the requested packets that is the middle bar showing the actual number of messages that arrived on the MAC layer. Lastly, the third bar at the most right represents the number of messages that are received.

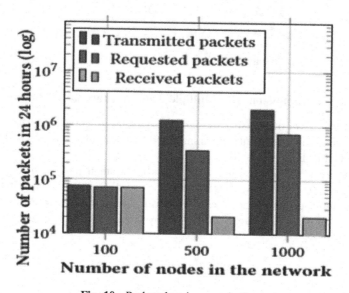

Fig. 10. Packets loss in scenario three

It is evident from Figs. 9 and 10, the transmissions with higher spreading factor effect the reliability of the LoRaWAN network, because it causes a lot of collisions; so, the nodes far from the gateway get stuck in a circle and could also hit their duty cycle limitation due to causing extra collisions while trying to increase the reliability of the channel.

5 Analysis

In this section, we have provided resource usages and analysis of the evaluation results of all scenarios as presented in Sect. 4.

The evaluation results show that the LoRaWAN networks are highly scalable, supporting of hundreds or thousands of end devices, both nodes and gateways. Nevertheless, it has shown in the results that when more nodes are added the performance drops i.e. packet error ratio increases. The reason this is because when more nodes are added so the nodes far away from gateway could not communicate and acknowledge packets from far away, and the packets usually drop reducing the network scalability. One possible way of diminishing the problem is to lower the traffic load on each device. We are focusing on the capacity of a single gateway for scalability analysis, where gateway capacity is defined by the number of devices that could be supported by the gateway at a predefined packet delivery ratio. Since the traffic load is similar for every number of devices in our simulations, so the results show a drop at reliability curves.

The Table 1 summarized memory usage and time in each simulation scenarios. Since all nodes sends messages in every two minutes for twenty-four hours. According to the statistics measurements, it is stated that small networks are simulated much faster than larger networks and that acknowledgements take much efforts as shown in Table 1. For latter cases, the requirements amplified expressively which is due to the acknowledgements and retransmissions in the network.

As presented in Table 2, the Packet Error Ratio (PER) is calculated to demonstrate the reliability and scalability of the IoT network scenarios. The PER increases with the nodes increased, and the distance increased in the network.

Table 1. Resource usages by each scenario in simulation

Number of nodes		100	500	1000
Scenario 1	Memory (MB)	465	2175	4355
	Time (s)	551	11756	47136
Scenario 2	Memory (MB)	517	2235	4378
	Time (s)	971	13773	48654
Scenario 3	Memory (MB)	1364	3635	10029
	Time (s)	1653	28894	115523

The throughput results in Tables 2 and 3 states the network reliability of the network. As observed from the Table 2, the throughput increases with increased number

of nodes, and the gain decreased in the network when there are less nodes. Lastly, the fairness among nodes measures the network performance.

Table 2. Packet error ratio, throughput and fairness measured in single cell scenario

Number of nodes	1000	3500
Packet error ratio (%)	7.5	20
Throughput (Kbps)	3	9.5
Fairness	1	0.977

Table 3. Packet error ratio, throughput and fairness measured in multiple cell scenario

Number of nodes	100	500	1000
Packet error ratio (%)	5	18	30
Throughput (Kbps)	0.25	1.45	2.55
Fairness	1	0.999	0.998

6 Conclusion

The paper has evaluated LoRaWAN standard to present scalability and performance in a number of IoT network scenarios. We have developed a LoRaWAN simulation model in ns-3 for long range communication. Using newly developed simulation for LoRaWAN, the paper has simulated three different scenarios which show that multiple gateways considerably improve the reliability of the network. The simulation also shows, even with multiple gateways, the PER significantly increases due to providing high reliability using acknowledgement for each packet in LoRaWAN based IoT network. The result of this simulation can add high value for designing LoRaWAN based IOT network for mission-critical application. Furthermore, the result can be used as a benchmark for improving reliability and scalability by developing new transmission techniques of LoRaWAN based IoT networks.

References

1. Ray, B., Chowdhury, M., Abawajy, J.: PUF-based secure checker protocol for networked RFID systems. In: 2014 IEEE Conference on Open Systems (ICOS), pp. 78–83. IEEE Computer Society Conference Publishing Services, Malaysia (2014)
2. Ray, B., Chowdhury, M., Abawajy, J.: Secure object tracking protocol for networked RFID systems. In: 16th ACIS International Conference on Software Engineering, Artificial Intelligence, Networking and Parallel/Distributed Computing (SNPD 2015), pp. 1–7. IEEE Computer Society, Japan (2015)
3. Raza, U., Kulkarni, P., Sooriyabandara, M.: Low power wide area networks: an overview. IEEE Commun. Surv. Tutor. **19**(2), 855–873 (2017)
4. The Things Industries (2019). https://www.thethingsnetwork.org/

5. Reynders, B., Wang, Q., Tuset-Peiro, P., Vilajosana, X., Pollin, S.: Improving reliability and scalability of lorawans through lightweight scheduling. IEEE Internet Things J. **5**(3), 1830–1842 (2018)
6. Tiurlikova, A., Stepanov, N., Mikhaylov, K.: Method of assigning spreading factor to improve the scalability of the LoRaWan wide area network. In: 2018 10th International Congress on Ultra Modern Telecommunications and Control Systems and Workshops (ICUMT), pp. 1–4. IEEE (2018)
7. Labs, L.: A comprehensive look at low power, wide area networks for internet of things engineers and decision makers. White Paper (2016). http://info.linklabs.com/lpwan-1. Accessed 15 July 2017
8. Yousuf, A.M., Rochester, E.M., Ghaderi, M.: A low-cost lorawan testbed for iot: implementation and measurements. In: 2018 IEEE 4th World Forum on Internet of Things (WF-IoT), pp. 361–366. IEEE (2018)
9. Bor, M., Vidler, J., Roedig, U.: LoRa for the Internet of Things. In: Proceedings of ACM EWSN, Graz, Austria, pp. 361–366 (2016)
10. Georgiou, O., Raza, U.: Low power wide area network analysis: can LoRa scale? IEEE Wirel. Commun. Lett. **6**(2), 162–165 (2017)
11. Mikhaylov, K., Petajajarvi, J., Janhunen, J.: On LoRaWAN scalability: empirical evaluation of susceptibility to inter-network interference. In: Proceedings of IEEE EuCNC, Oulu, Finland, pp. 1–6 (2017)
12. Haxhibeqiri, J., Van den Abeele, F., Moerman, I., Hoebeke, J.: LoRa scalability: a simulation model based on interference measurements. Sensors **17**(6), 1193 (2017)
13. Reynders, B., Meert, W., Pollin, S.: Range and coexistence analysis of long range unlicensed communication. In: Proceedings of IEEE ICT, Thessaloniki, Greece, pp. 1–6 (2016)
14. Pop, A.-I., Raza, U., Kulkarni, P., Sooriyabandara, M.: Does bidirectional traffic do more harm than good in lorawan based LPWA networks?. In: Proceedings of IEEE GLOBECOM, Singapore, pp. 1–6 (2017)
15. Reynders, B., Meert, W., Pollin, S.: Power and spreading factor control in low power wide area networks. In: Proceedings of IEEE ICC, Paris, France, pp. 1–6 (2017)
16. Adelantado, F., et al.: Understanding the limits of LoRaWan. IEEE Commun. Mag. **55**(9), 34–40 (2017)
17. Tuset-Peiro, P., Vazquez-Gallego, F., Alonso-Zarate, J., Alonso, L., Vilajosana, X.: LPDQ: a self-scheduled TDMA MAC protocol for onehop dynamic low-power wireless networks. Pervasive Mobile Comput. **20**, 84–99 (2015)
18. Zhang, K., Marchiori, A.: Crowdsourcing low-power wide-area IoT networks. In: Proceedings of IEEE International Conference on Pervasive Computing and Communications (PerCom), Kailua-Kona, HI, USA, pp. 41–49 (2017)
19. Weightless-SIG: Neul's Weightless-N (2015). http://www.weightless.org
20. Ingenu (2017). http://ingenu.com
21. LoRaWAN: The Things Network. https://www.thethingsnetwork.org/docs/lorawan/. Accessed 27 Apr 2019
22. Neumann, P., Montavont, J., Noël, T.: Indoor deployment of low-power wide area networks (LPWAN): a LoRaWAN case study. In: 2016 IEEE 12th International Conference on Wireless and Mobile Computing, Networking and Communications (WiMob), pp. 1–8. IEEE (2016)
23. https://www.nsnam.org/
24. Reynders, B., Wang, Q., Pollin, S.: A LoRaWAN module for ns-3 (2018)

An Efficient Dynamic Group-Based Batch Verification Scheme for Vehicular Sensor Networks

Liming Jiang[1], Yan Wang[2], Jiajun Tian[2], Frank Jiang[3(✉)],
Nick Patterson[3], and Robin Doss[3]

[1] School of Computer Science and Technology, Hunan University of Science
and Technology, Xiangtan 411100, Hunan, China
[2] School of Computer Science and Technology, University of South China,
Hengyang 421001, Hunan, China
[3] School of Information Technology, Deakin University,
Geelong 3216, Australia
Frank.Jiang@deakin.edu.au

Abstract. The batch verification methods can relieve the bottleneck problem of vehicular authentication efficiency to accelerate the authentication speed during the group construction phase. However, in current batch verification methods, there exist some deficiencies such as the authentication methods applied in the actual scenario are sensitive to the group initialization and dynamic construction process and the intra-group communication efficiency have not been improved due to lack of effective dynamic group management. This paper proposes a fast and secure batch verification scheme based on dynamic group management and certificateless public-key cryptography. First, the two-way authentication between the proxy vehicle and the roadside unit (RSU) is established, and then the batch verification method relying on RSU-assisted are proposed based on dynamic management of the intra-group vehicles. Our method fully considers the security requirements of the group while solving certification efficiency. Theoretical analysis and simulation experiments show that compared with other existing solutions, the verification scheme in this paper reduces the computational delay and transmission overhead whilst increasing robustness in dynamic environments.

Keywords: VANETs · Dynamic group · Batch verification · RSU-assisted

1 Introduction

Vehicular Ad Hoc Networks (VANETs) are a mobile self-organizing network designed to enable vehicles to communicate with each other, vehicle-to-vehicle (V2V) communications, communication between vehicles and vehicles (V2V), vehicle-to-infrastructure (V2I) communication, or a hybrid of the two, called vehicle-to-vehicle-to-infrastructure (V2V2I) communication. The VANETs are generally composed of vehicles equipped with on-board units (OBUs), the roadside unit (RSUs) and a Trust Authority (TA). As the main component of the intelligent transportation system

© Springer Nature Switzerland AG 2019
R. Doss et al. (Eds.): FNSS 2019, CCIS 1113, pp. 52–64, 2019.
https://doi.org/10.1007/978-3-030-34353-8_4

(ITS), VANETs can improve traffic efficiency and driving safety while reducing or avoiding traffic accidents and providing entertainment and information services using communication between V2V and V2I, as well as accessibility to the Internet. In recent years, VANETs has entered into a stage of rapid development with huge market potential and social benefits.

Security and privacy protection are two important issues which will determine whether VANETs can be accepted, promoted and applied for real-world uses [1–4]. First, the authentication, as well as the assurance of integrity and traceability of the information provided by some vehicles, is an effective way to solve VANETs security and then, another aspect which must be considered is privacy protection, meaning that the private information of users such as the driver's name, the license plate, speed and travelling routes should be protected. Lastly, because of the fast-moving nature of vehicles and the small range of RSU communication coverage, the computational and communication costs incurred by the frequency of authentication between vehicles and the nearest RSUs, will also become a bottleneck problem that hinders the widespread deployment of VANETs.

Compared to improving the one-on-one authentication efficiency for the vehicle and the RSU, the batch verification method, in which one-to-many verification is done for a group proxy and the RSU, will allow the better handling of the efficiency bottleneck caused by the RSU coverage and processing delay, which actually has received attention from many domestic and foreign scholars [5–7]. However, the existing group-based batch verification methods are focused on the implementation of the verification process and neglected the impact of the dynamic group management and maintenance on the applicability and efficiency of the verification method to a certain degree. This is mainly reflected in the follow points: (1) The group construction method is too simple and it is difficult to meet the requirements of vehicles' rapid mobility and topological dynamics of the VANETs (2) The lack of effective group initialization, dynamic construction and update methods, which will directly affect the applicability and efficiency of verification methods in practical scenarios.

Based on the existing group-based batch verification methods, the characteristics of VANETs such as being self-organizing, fast-moving and volatile topological, as well as the security requirement of the group that supports batch verification, are considered fully. A dynamic group construction and updating scheme are proposed in this paper to meet the needs of batch authentication. On this basis, a batch verification method based on the dynamic group is proposed to improve the efficiency. Finally, the applicability and efficiency of the proposed scheme are analyzed, and the performance is compared with other methods by simulating.

2 Related Work

To deal with the privacy protection of VANETs, many authentication schemes have been proposed. The literature detailed in [8–11] adopt the identity-based public-key encryption system which reduces the computational overhead and transmission overhead compared with traditional public-key cryptosystems. Literature [8] proposes a batch verification method that can accelerate the vehicular authentication speed, but the

privacy leakage has not been resolved. The literature [9–11] proposes an effective privacy protection scheme, but the attacker can easily trace back to the real identity of the vehicle who is sending the information. Literature [10] proposes a key management scheme for batch verification that can speed up the vehicular authentication speed but also lacks consideration of revocability. The literature [11] satisfies the feasibility of batch certification, but it is not a two-way authentication method. It only considers vehicular authentication and does not consider vehicle-initiated authentication of RSUs. Also, all the above methods have a key escrow issue, so that the private key generation center can obtain the communication content between vehicles, thus divulging the vehicle's privacy, which is also not allowed by the vehicle user. Therefore, a new method is needed to solve the above problem.

As to the group construction in VANETs, the literature [5] divides the network area into sections according to geographical locations. All the vehicles travelling in the same area are equivalent to being in the same group and only initiate intra-group communication. In this scenario, the vehicles selection in the group is inflexible, and the number of vehicles in different regions is often very different. In a certain period of time, there may be many vehicles in the same area, which will make the key management very inconvenient. Subsequently, [6] proposed that a vehicle can be flexibly formed into a group, but the solution allows the trusted authority to download its own master private key to each registered vehicle. Although the hardware installed on the vehicle can avoid global security failure, it brought about a significant increase in hardware costs, and would likely be unacceptable for most vehicle users.

For the problem that the existing solutions cannot fully satisfy is the authentication requirements in VANETs. This paper proposes a fast and secure batch verification scheme based on dynamic group management and certificateless public-key cryptography. While ensuring communication safety, the computational communication overhead of vehicles and RSUs is reduced effectively.

3 Dynamic Group Batch Verification

Once the vehicle entered the RSU communication range, it would get a temporary pseudonym from the RSU. If the vehicle is the first and the only one having entered the RSU communication range, and have been successfully verified by the RSU, it will create a dynamic group and become the group proxy. The group proxy vehicle broadcasts the advertising message about the group and collects the surrounding vehicles' verification information. The collected information is sent to the RSU in batches, so the time for information collection is significantly reduced.

3.1 System Initialization

Let G_1 be the q-order cyclic additive group generated by P, and G_2 be a cyclic multiplicative group with the same order as G_1. $G_1 \times G_2 \rightarrow G_2$ is a bilinear map. TA selects a random number $S(S \in Z_q^*)$ as its private key, and the corresponding public-key, PK can be calculated as $PK = s \cdot P$, Where s is private to TA. Every vehicle and RSU must download the system's common parameters $\{G_1, G_2, q, P, PK, H(\cdot), H_1(\cdot)\}$

from the TA at the beginning. Among the parameters, $H(\cdot)$ is a one-way hash function and $H_1(\cdot)$ is a hash function mapped to a point. In addition, each RSU also obtains a signature $Sig_R(PKR)$, which is signed by TA on the RSU's public key (PKR) with its private key. Each vehicle sends its own unique real identity RID_i to the TA through the existing secure channel, then the partial private key of the vehicle is calculated by the TA and the value can be expressed as $Y_i = s \cdot RID_i$. According to the difficult problem of ECDLP, it can be seen that even if the vehicle knows Y_i and RID_i, the TA private keys cannot be found. The vehicle then generates a unique private key (R_i, Y_i), Where R_i is a random number generated by the vehicle V_i.

3.2 Mutual Authentication Between Group Proxy Vehicle and RSU

In our method, there are two situations in which the vehicle would be selected as the group proxy vehicle: one is the vehicle who first requires services and lies within the RSU communication range and can complete mutual authentications with the RSU. The other is that the proxy vehicle leaves the RSU communication range and the RSU randomly selects an intra-group vehicle to become the new group proxy vehicle.

The vehicular networks communication process based on dynamic group batch verification is established as shown in Fig. 1. The proxy vehicle V_L obtains some information from RSU, such as broadcast message $Sig_R(PKR)$ and the public key PKR of the RSU. Once the RSU passes V_L's certification, the message $(PID_L, ID_m, PK_L, Enc_{PKR}(N_1), T_1, Sig_L)$ is sent to the RSU by V_L. After V_L is verified by RSU, the group communication key and session key which are generated by RSU would be transferred to V_L.

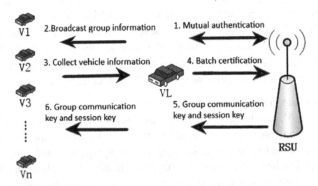

Fig. 1. Dynamic group batch verification flowchart

3.3 RSU-Assisted Batch Verification

Suppose the group with n vehicles excluding the proxy vehicle and needs to collect vehicle information $(PID_i, ID_m, PK_i, Sig_i)$, where $Sig_i = r_i \cdot Y_i \cdot A_i + r_i \cdot B_j, A_i = H(PID_i^1, PID_i^2, PK_i, ID_m), B_j = H_1(ID_m)$. The proxy vehicle V_L aggregates the

signatures from n messages generated by the intra-group vehicles, and the formula is expressed as $\text{sig} = \sum_{i=1}^{n} \text{Sig}_i$. The aggregation process is shown in Fig. 2.

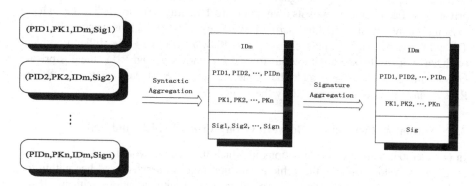

Fig. 2. Syntax aggregation and signature aggregation

3.4 RSU-Assisted Dynamic Group Management

Let $V_0, V_1, V_2, \ldots, V_n$ denote n + 1 vehicles respectively, where V_L denotes the group proxy node, and the dynamic group management process is described as follows:

(1) Dynamic group initialization

Once vehicle V_i enters the RSU_j communication coverage, V_i generated a random number N_1 and message sequence $(\text{PID}_L, \text{ID}_m, \text{PK}_L, \text{Enc}_{\text{PKR}}(N_1), T_1\text{Sig}_L)$ which is sent to RSU_j if no group broadcast information is received from other group proxy. After the successful mutual authentication with the RSU_j, the vehicle V_i becomes a new group proxy. Firstly, RSU_j decrypt $\text{Enc}_{\text{PKR}}(N_1)$ to obtain N_1, and generates a random number N_2, then RSU_j calculates the session key K and sends the message $(\text{Enc}_{\text{PK}_i}(N_2), T_2, \text{Enc}_{K_i}(\text{Mes}))$ to the group proxy. At last, the group proxy decrypt $(\text{Enc}_{\text{PK}_i}(N_2))$ to obtains the random number N_2 and calculates the session key K.

(2) Dynamic group construction

The proxy vehicle V_L sends the group broadcast messages and the surrounding vehicle that need to communicate with RSU would reply to V_L. V_L checks the size of the group and the current capacity before collecting the replying messages of surrounding vehicles that need to be communicated with RSU. If the latter is less than the former, then V_L would accept the join request, otherwise V_L will refuse the node to join and no longer send the group broadcast message until the capacity is reduced. The message format sent by the vehicle V_i is denoted as $(\text{PID}_i, \text{PK}_i, \text{Agr}, T_4, \sigma_i)$. V_L sends the received n pieces of messages to the RSU_j through the session key K. RSU_j performs batch verification by the following formula (1):

$$\begin{cases} A = e\left(H_2(\text{Agr}), \sum_{i=1}^{n} PK_i \right) \\ B = e\left(\sum_{i=1}^{n} PID_i^1 h(W_i), PK \right) \\ e\left(\sum_{i=1}^{n} \sigma_i, P \right) = AB \end{cases} \tag{1}$$

After the batch certification having been passed, RSU_j uploads the group vehicles' pseudonym information and public key information to the TA. Firstly, TA generates the service key $K_i (i = 1, 2, \ldots, n)$ corresponding to the n vehicles and calculates the group communication key $Sig_s(K_G)$ for all the vehicles within the same group. Then the message denoted as $V = (Enc_{PK_1}(Sig_s(K_G), K_1), Enc_{PK_2}(Sig_s(K_G), K_2), \ldots, Enc_{PK_{n-1}}(Sig_s(K_G), K_{n-1}))$, is encrypted by the CA and broadcast within the group. If the certification is not passed, the untrustworthy vehicles must exclude and upload the legal vehicle information to TA again.

(3) Ordinary vehicle leaves
When the ordinary vehicle $V_i (i = 1, 2, \ldots, n)$ left the group, it needs to send messages denoted as $Enc_{K_i}(U, Sig_{sk_i}(U))$ and $U = (PID_i, PK_i, EG)$ to the RSU_j after receiving the message, RSU_j decrypt the U and $Sig_{sk_i}(U)$ with the corresponding service key. RSU_j generates a random number denoted as d and sends it to other members of the group to update the group key, RSU_j broadcast the message denoted as $V' = (Enc_{PK_j}(d))$, where $j = 1, 2, \ldots, n (j \neq i)$. The vehicle nodes in the group decrypted V' with their respective private keys to obtain the random number d, and then a new intra-group communication key $K_G' = K_G \cdot d$ is generated, so as to exclude vehicles V_i from the group.

(4) Group proxy vehicle leaves
When the group proxy vehicle V_L left the dynamic group, it needs to send a message $\{ Enc_{K_L}(U, Sig_{sk_L}(U)), U(PID_L, PK_L, EG) \}$ to the RSU_j before leaving the group. After the RSU_j receives the message, it uses the corresponding service key to decrypt the encrypted message and obtain U and $Sig_{sk_L}(U)$ from the computing results. RSU_j randomly selects the intra-group vehicle $V_i (i \neq L)$ as the group proxy and encrypts the intra-group vehicles' information $\{ Enc_{K_i}(L, Sig_{pk_i}(L) \}$ with the corresponding service password K_i, where L is a message can be expressed as $((PID_1, PK_1), (PID_2, PK_2), \ldots, (PID_{n-2}, PK_{n-2}))$, and then sent them to the vehicle $V_i (i \neq L)$, excluding the proxy vehicle whether it is original or current. RSU_j uses the secret key K_G shared by the group to encrypt the message including the new proxy of the vehicle's information and broadcasts it within the group. RSU_j generates a random number d, construct message $V'' = (Enc_{PK_i}(d))$ and broadcasts it within the group to perform the group key update. Vehicle nodes in the group are decrypted with their respective private keys to obtain a random number d', then generate a new intra-group communication key $K_G' = (K_G \cdot d)$, thereby excluding the vehicle V_L from the group.

3.5   Security and Performance Analysis

(1) Security analysis

Correctness and completeness of information: messages collected by the proxy vehicle is denoted as $(\mathrm{PID_i}, \mathrm{PK_i}, \mathrm{ID_m}, \mathrm{Sig_i})$, where $\mathrm{Sig_i} = r_i \cdot Y_i \cdot A_i + r_i \cdot B_j$, $A_i = H(\mathrm{PID_i^1}, \mathrm{PID_i^2}, \mathrm{PK_i}, \mathrm{ID_m})$, $B_j = H_1(\mathrm{ID_m})$. Because r_i is a random private key generated by the vehicle V_i, and Y_i is a part of the private key issued by TA, thus the signature is unforgeable. The proxy vehicle judges the correctness of the equation as well as the verification information having been modified through verifying the signature. If the verification fails, the message is discarded to ensure the correctness and completeness of the message.

Privacy and anonymity: The temporary pseudonym of the vehicle is denote as $\mathrm{PID_i} = (\mathrm{PID_i^1}, \mathrm{PID_i^2})$ where $\mathrm{PID_i^1} = r_i \cdot \mathrm{RID_i}, \mathrm{PID_i^2} = \mathrm{RID_i} \oplus H_1(r_i \cdot \mathrm{PK})$. As long as the vehicle first enters or re-enters the RSU communication range, it will receive a pseudonym certificate issued by RSU to ensure the continuous updating of the pseudonym. At the same time, since r_i is randomly generated by the vehicle and has unpredictability, the attacker could not predict the vehicle's pseudonym, or even judge whether the two messages in the group were from the same vehicle. This effectively avoided the risk of the vehicle being tracked. In our method, the pseudonym of the vehicle is composed of two parts. Once the attacker breaks through the TA, the pseudonym of the vehicle still cannot be obtained, so the privacy of the vehicle can be better protected.

Non-repudiation: VANETs privacy protection is a conditional privacy protection. Under normal circumstances, the vehicle sends messages anonymously, and TA can trace the true identity of the message sender while disputes are still occurring. Through joint use of its' own private key s, the temporary public key $\mathrm{PK_i}$ uploaded by the vehicle V_i, the temporary pseudonym $\mathrm{PID_i} = (\mathrm{PID_i^1}, \mathrm{PID_i^2})$ and the $H_1(\cdot)$ which is hash function mapped to point and obtained from the TA at the time of vehicle initialization, the TA can calculate and obtain the true identity of the vehicle. The calculation derivation of formula (2) is as follows:

$$
\begin{aligned}
&\mathrm{PID_i^2} \oplus H_1(s \cdot \mathrm{PK_i}) \\
&= \mathrm{RID_i} \oplus H_1(r_i \cdot \mathrm{PK}) \oplus H_1(s \cdot \mathrm{PK_i}) \\
&= \mathrm{RID_i} \oplus H_1(r_i \cdot \mathrm{PK}) \oplus H_1(s \cdot r_i \cdot P) \\
&= \mathrm{RID_i} \oplus H_1(r_i \cdot \mathrm{PK}) \oplus H_1(r_i \cdot \mathrm{PK}) \\
&= \mathrm{RID_i}
\end{aligned}
\tag{2}
$$

(2) Performance analysis

We can see that the vehicle's pseudonym refresh is triggered by the RSU, but it is generated automatically by the vehicle, so it does not need a lot of pseudonym storage. If the vehicle wants to join the group communication, it must send a group request message to the group proxy vehicle. The proxy is responsible for collecting all the

request messages and sending to the base station for verification. The RSU performs batch verification by using formula (1). Compared with other single verification methods, our method can effectively optimize the system performances such as communication time and computational overhead, so it is of great significance in large-scale VANET deployment. When an ordinary vehicle leaves the group, it only needs to send one random number to the intra-group member to update the K_G. When the group proxy leaves, it only needs to reselect the group proxy and sends all the group members' information to the proxy and sends a random number to the intra-group members to update the K_G, the calculation and storage overhead is saved effectively.

Verification overhead: It is assumed that the time cost for calculating a bilinear pairing operation is T_{par}, the time required for calculating a hash operation mapped to a point is T_{mtp}, the time for calculating a point multiplication operation is T_{mul}, and the other operation time is ignored. The verification overhead is shown in Table 1 below, where DGBS refers to the scheme of this article, and IBV refers to the method proposed in literature [7]. The operation T_{mtp} represents the time required to calculate a hash operation mapped to a point, and T_{mul} represents the time required to calculate a dot multiplication operation, T_{par} represents the time required to calculate a bilinear mapping pair.

Table 1. Vehicle verification calculation overhead

Scheme	One vehicle certification	N vehicle certification
IBV	$3T_{par} + T_{mtp} + T_{mul}$	$3nT_{par} + nT_{mtp} + nT_{mul}$
DGBS	$3T_{par} + T_{mtp} + T_{mul}$	$3T_{par} + T_{mtp} + nT_{mul}$

Signature generation overhead: The vehicle is allowed to join the dynamic group after an authentication request message having passed the verification of RSU. Whether the message is sent from the ordinary vehicle to the group proxy vehicle or from group proxy vehicle to the RSU, it needs to be signed before sending. The calculation cost of vehicular signature generation is shown in Table 2 below:

Table 2. Vehicle signature generation calculation overhead

Scheme	One vehicle generated signature
IBV	$2T_{mtp} + 5T_{mul}$
DGBS	$T_{mtp} + 5T_{mul}$

Signature verification overhead: After the receiver receives the signed message, it needs to verify whether the signature is correct or not. If yes, the receiver accepts the message. Otherwise discarded. The computation cost of vehicular signature verification is shown in Table 3 below:

Table 3. Vehicle verification signature calculation overhead

Scheme	One vehicle verification signature	N vehicle verification signature
IBV	$3T_{par} + T_{mtp} + 2T_{mul}$	$3T_{par} + nT_{mtp} + 2nT_{mul}$
DGBS	$2T_{par} + T_{mtp} + 2T_{mul}$	$2nT_{par} + nT_{mtp} + 2nT_{mul}$

Transmission cost: In our method, the length of a message mainly depends on the pseudonym and signature. Assume that the signature has V_s bytes and the pseudonym has V_p bytes. Other overheads are ignored. The transmission overhead is shown in Table 4 below:

Table 4. Vehicle authentication transmission costs

Scheme	Send one message	Send n messages
IBV	$V_s + V_p$ byte	$(V_s + V_p)n$ byte
DGBS	$V_s + V_p$ byte	$V_s + nV_p$ byte

As the number of authentication messages received by the RSU increases, the amount of information transmitted over the entire network increases accordingly. The comparison chart of the communication expenses between the IBV scheme and our scheme is shown in Fig. 3 below. From the experimental results, our method is better than IBV. As shown in Table 4, we can see that in the case of transmitting one authentication message, the transmission overhead of the two methods is the same, but when transmitting 'n' aggregate messages at a time, the method utilized in this paper only requires the RSU to perform one signature operation, so reduced transmission overhead.

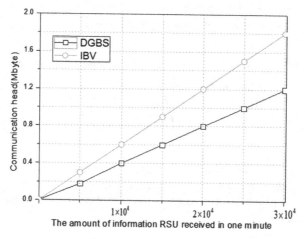

Fig. 3. Communication overhead of RSU receiving information in one minute

4   Simulation and Experiment Analysis

In this paper, the simulation was made on the traffic simulator VanetMobiSim and network emulator NS2, both are free and open sourced. VanetMobiSim generates a mobile topology for the vehicle based on the configuration file, and NS2 simulates the network by using the mobile topology generated by VanetMobiSim. Experimental simulation parameters and environmental settings are similar or identical to existing literature [7, 9] (Table 5).

Table 5. Simulation parameters

Simulation parameters	Value
Road length	1000 m
Simulation period	20 s
Pause period	0 s
MAC protocol	802.11p
Channel bandwidth	6 Mbps
Transmission radius	300 m

The experiment mainly analyzes the impact of the number of vehicles and RSUs in the system on verification delay and authentication success rate. The average verification delay of messages defined in this paper is expressed as follows:

$$AverageMessageDlay = \frac{1}{N}\frac{1}{M}\sum_{i=1}^{N}\sum_{m=1}^{M}\left(T_{verif}^{i-m-rsu} + T_{sign}^{i-m} + T_{trans}^{i-m-rsu}\right) \quad (3)$$

Where M is the amount of messages sent by the vehicle V_i, N is the number of vehicles, $T_{verif}^{i-m-rsu}$ denotes the time required for the RSU to verify the message m sent from the vehicle V_i, T_{sign}^{i-m} represents the time it takes for the vehicle to sign the message m, and $T_{trans}^{i-m-rsu}$ denotes the transmission time of the message when vehicle V_i sends message m to RSU.

Due to the fast mobility of the vehicles, if the verification messages cannot be validated quickly, the messages will be discarded. The successful verification ratio (SRC) is defined as the ratio of the number of verification messages submitted by successfully verified vehicles to the number of verification messages submitted by all vehicles within a certain period.

$$SRC = \begin{cases} 1 & \frac{d}{T.V.N} \geq 1 \\ \frac{d}{T.V.N}, & \text{otherwise} \end{cases} \quad (4)$$

Where d indicates how long it takes for these vehicles entering the RSU to complete the certification, T represents the sum of calculation time when completing an authentication request, and v indicates the speed of the vehicle.

Figure 4 mainly describes the impact of the vehicle's travel speed on the average message delay. It can be seen that the mean time delay of our method is the shortest when the average vehicle speed increases from 5 m/s to 40 m/s. The AKA [9] and IBV [7] methods are affected by the speed and packet length, so the average delay is relatively large as shown in Fig. 6 below.

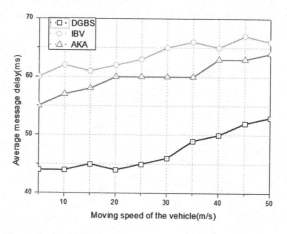

Fig. 4. Impact of vehicle's moving speed versus average message delay

As can be seen from Fig. 5, when the number of vehicles is 20, the average message delays of the AKA method and the IBV method are about 60 ms and 63 ms respectively, and the average message delay of our method is about 44 ms. When the number of vehicles reaches 100, the average message delay for our method is approximately 101 ms, while AKA and IBV are approximately 195 ms and 197 ms, respectively. Therefore, in these three methods, the average message delay in our method increases the slowest, as the number of vehicles increases.

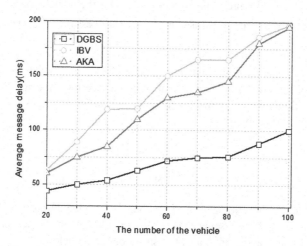

Fig. 5. Average time delay of different numbers of RSU group information

Figure 6 shows the success rate of vehicle batch certification in various methods, where D = 30 m and v = 25 m/s. It can be seen that after the number of vehicles reached 250 vehicles and 300 vehicles respectively, the RSUs in the AKA and IBV methods were unable to respond to partial verification information in a timely manner. As the number of vehicles increases, this phenomenon will become more and more serious. In our method, only after the number of vehicles reached 600, the RSU began to fail to respond to part of the vehicles in a timely manner, indicating that this method is highly robust.

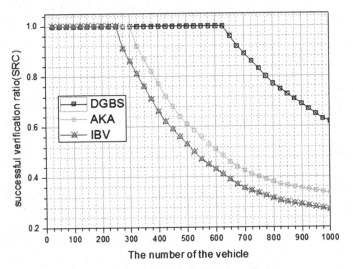

Fig. 6. Probability of successful batch verification

5   Conclusion

In this paper, we analyzed current batch verification methods for VANETs and showed that these methods have some bottleneck problems regarding vehicular authentication efficiency caused by the lack of effective dynamic management. Moreover, we suggested an improved dynamic group-based batch verification scheme to reduces the computational delay and transmission overhead whilst increasing robustness in dynamic environments. Our scheme consists of three parts: mutual authentication of group proxy vehicle and RSU, RSU-assisted batch verification, and RSU-assisted dynamic group management. The simulation expressed the efficiency and practicality of our scheme.

References

1. Qu, F., Wu, Z., Wang, F.-Y., Cho, W.: A security and privacy review of VANETs. IEEE Trans. Intell. Transp. Syst. **16**(6), 2985–2996 (2015)

2. Al-Sultan, S., Al-Doori, M.M., Al-Bayatti, A.H., Zedan, H.: A comprehensive survey on vehicular ad hoc network. J. Netw. Comput. Appl. **37**, 380–3923 (2014)
3. Lu, R., Lin, X., Liang, X., Shen, X.: A dynamic privacy-preserving key management scheme for location-based services in VANETs. IEEE Trans. Intell. Transp. Syst. **13**(1), 127–139 (2012)
4. Xiong, H., Chen, Z., Li, F.: Efficient and multi-level privacy-preserving communication protocol for VANET. Comput. Electr. Eng. **38**(3), 573–581 (2012)
5. Raya, M, Aziz, A, Hubaux, J.P.: Efficient secure aggregation in VANETs. In: Proceedings of International Workshop on Vehicular Ad Hoc Networks, New York, pp. 67–75 (2006)
6. Chim, T.W., Yiu, S.M., Lucas, C.K., et al.: SPECS: secure and privacy enhancing communications schemes for VANETs. Ad Hoc Netw. **9**, 189–203 (2011)
7. Zhang, C., Lu, R., Lin, X., et al.: An efficient identity based batch verification scheme for vehicular sensor networks. In: The 27th IEEE Conference on Computer Communications, INFOCOM, Phoenix, AZ, pp. 246–250 (2008)
8. Lu, R., Lin, X., Zhu, H., et al.: ECPP: efficient conditional privacy preservation protocol for secure vehicular communications. In: Proceedings of IEEE INFOCOM, Phoenix, pp. 1229–1237 (2008)
9. Yang, J.H., Chang, C.C.: An ID-based remote mutual authentication with key agreement scheme for mobile devices on elliptic curve cryptosystem. Comput. Secur. **28**, 138–143 (2009)
10. Huang, J.L., Yeh, L.Y., Chien, H.Y.: ABAKA: an anonymous batch authenticated and key agreement scheme for value-added services in vehicular ad hoc networks. IEEE Trans. Veh. Technol. **60**, 248–262 (2011)
11. Shim, K.A.: CPAS: an efficient conditional privacy-preserving authentication scheme for vehicular sensor networks. IEEE Trans. Veh. Technol. **61**, 1874–1883 (2012)
12. Sun, Y., Lu, R., Lin, X., et al.: An efficient pseudonymous authentication scheme with strong privacy preservation for vehicular communications. IEEE Trans. Veh. Technol. **59**, 3589–3603 (2010)
13. Liu, J.K., Yuen, T.H., Au, M.H., Susilo, W.: Improvements on an authentication scheme for vehicular sensor networks. Expert Syst. Appl. **41**(5), 2559–2564 (2014)
14. Liu, Y., Wang, L., Chen, H.-H.: Message authentication using proxy vehicles in vehicular ad hoc networks. IEEE Trans. Veh. Technol. **64**(8), 3697–3710 (2015)
15. Zhang, C., Lin, X., Lu, R., Ho, P.H.: RAISE: an efficient RSU-aided message authentication scheme in vehicular communication networks. In: ICC 2008, IEEE International Conference on Communications, pp. 1451–1457. IEEE (2008)
16. Shao, J., Lin, X., Lu, R., Zuo, C.: A threshold anonymous authentication protocol for VANETs. IEEE Trans. Veh. Technol. **65**(3), 1711–1720 (2016)
17. Chuang, M.-C., Lee, J.-F.: TEAM: trust-extended authentication mechanism for vehicular ad hoc networks. IEEE Syst. J. **8**(3), 749–758 (2014)

MetaCom: Profiling Meta Data to Detect Compromised Accounts in Online Social Networks

Ravneet Kaur[✉], Sarbjeet Singh, and Harish Kumar

University Institute of Engineering and Technology, Panjab University,
Chandigarh, India
ravneets48@gmail.com, {sarbjeet,harishk}@pu.ac.in

Abstract. Social networks have become the center of research and its increasing popularity has also led to its misuse by a number of malicious users. In order to conduct various malicious activities on the online platforms, malevolent users rely on spam, fake or compromised accounts to disseminate their illegitimate information. This paper addresses the detection of compromised accounts so that the concerned user can take the necessary action to mitigate the effect of compromise. Unlike most of the existing techniques where text based features are used to address the problem, this research examines the efficiency of meta data information associated with each text in detecting the compromised accounts. Secondly, we have studied the problem from both unary as well as binary classification perspectives where efficiency of respective machine learning classifiers have been analyzed on the basis of different evaluation metrics. Amongst five binary classifiers, Random Forest attained highest efficiency achieving 92.66% F-score. On the other hand, with one class classifiers, OCC-SVM with rbf kernel attained maximum performance with an average F-score of 72.72%.

Keywords: Compromised accounts · Binary classification · Unary classification · Online social networks

1 Introduction

In this era, online social networks have become an essential part of everyone's life bringing a revolution in their communication behavior. This immense popularity is evident from the fact that social networks have been ranked as one of the top most communication means where people with different backgrounds interact and share information among themselves [10]. But the concept of social networks which started only as a general framework for social contacts also brought with it numerous problems. It is commonly observed that cyber-criminals such as spammers, fraudsters and other unauthenticated users misuse the platforms for illegitimate purpose such as disseminating spam content, spreading unsolicited information, populating deceitful advertisements and many more [2,11,28].

© Springer Nature Switzerland AG 2019
R. Doss et al. (Eds.): FNSS 2019, CCIS 1113, pp. 65–80, 2019.
https://doi.org/10.1007/978-3-030-34353-8_5

In order to detect and minimize the effect of malicious activities, academicians and practitioners have proposed numerous techniques to detect spam and fake accounts but such techniques fail to distinguish much between fake and compromised accounts. Fake accounts are created just for the dissemination of spam and malicious content and once suspected the best solution is to delete them. On the other hand, with compromised accounts, original legitimate users lose access to their accounts and hence detection of such accounts also involves credential recovery as well as return of the authentic access of accounts to the genuine users.

Service providers of popular social networks rely only on point-of-entry based approaches such as IP geolocation based authentication at the time of login which seems to be insufficient. After the legitimate login into an account, the authenticated session can be easily commandeered by an intruder through various cookie stealing and session hijacking attacks. Hence, the standard static login authentication applied only once and that too at the start of the session should be supplemented with other reliable practices that could be applied unobtrusively at the back end. This research work aims to build such an unobtrusive behavioral modeling technique that profiles various meta data characteristics of each user and detects any abnormality in pattern to mark the point of compromise.

This research provides a three-fold contribution as follows:

- Firstly, it helps in building behavioral profile of a user using various meta-data features.
- Secondly, efficiency of meta data features is examined to determine the applicability in compromised account detection.
- Thirdly, problem has been studied from unary as well as binary classification perspectives and hence, efficiency of respective machine learning classifiers have been examined.

The proposed approach works on the idea that regular activities followed by the users could be easily modeled and whenever the account of an individual gets compromised, there is a perceptible change in the behavioral pattern of the user (this assumption is further confirmed by the experiments conducted). Behavioral models (profiles) pertaining to different modalities are constructed to capture the historical pattern of a user. Each model represents a particular feature of the message (e.g. a particular language in which a message is written or the time at which it was sent). Very general activity patterns are profiled and modeled which helps the approach to be deployed on any social network platform. Any future message that differ from the typical behavior of a user may denote a point of compromise.

Unlike most of the existing works that rely on the profiling of only textual features for the detection of compromised accounts [5–7], MetaCom addresses the problem by using the meta data information (such as language, time, source etc.) associated with each message. From the existing literature it is observed that the use of text based features have attained a good performance at detecting compromised accounts but nonetheless it is hypothesized that utilizing the side

(meta) information may help further improve the detection performance as the users not adhering to their textual profiles may abide by the tweeting patterns.

Overall objective of this research is to propose and develop a meta data profiling based technique that helps in detecting the compromised accounts in real-time by analyzing the change in behavior patterns.

Rest of the paper is structured as follows. Section 2 briefly discusses the related work done in the domain of compromised account detection. Section 3 presents an overview of the proposed methodology adopted followed by the discussion of experimental results in Sect. 4. Finally, the paper has been concluded in Sect. 5.

2 Related Work

In this section some of the existing works on the detection of compromised accounts have been summarized.

Unlike other related domains such as spam and fake account detection, work on detecting compromised accounts is still in its infancy. Researchers have started developing techniques to detect and mitigate the effect of compromised accounts only lately. Techniques developed to handle the problem either focus on the use of threshold based decision making or supervised/unsupervised learning.

Researchers such as Brocardo et al. [5–7], Igawa et al. [14], Jenny et al. [21], Barbon et al. [4], Kaur et al. [17] utilized a text mining approach in their respective works where efficiency of various textual features was examined to detect compromised accounts. Similarly, some of the research works have been carried out utilizing different message characteristic features. Egele et al. [8] proposed an approach named COMPA in which various features such as source, language, time, topic, links and proximilty were utilized. COMPA used an unsupervised clustering approach to detect the compromised accounts tracking the spam campaigns being conducted. Work was further extended by researchers such as Egele et al. [9], Trang et al. [29] to include some more case studies and focus on the detection of compromisation of high profile accounts. Nauta et al. [22,23] also deployed some meta data features for the detection of compromised accounts. Various filter and wrapper methods were used for feature selection followed by the application of two supervised classification algorithms namely, J48 and SMO. Unlike other works, VandDam et al. [30] used both text as well as meta features for the analysis of compromised accounts categorizing the intention behind compromisation into spread of spam, criticism, false proclamation and praises etc.

Apart from the textual and meta data characteristics, techniques based on temporal analysis [12,16,20,31], graph structure analysis [13,19] or introversial/extroversial characteristics [27] have also been developed to detect compromised accounts. This research aim to develop MetaCom, a meta data based profiling technique to handle the problem of compromised account detection from supervised machine learning perspective.

3   Proposed Work

This section discusses the methodology adopted for the detection of compromised accounts in online social networks. A supervised learning approach has been adopted which involves the communal steps beginning from data collection, feature extraction and selection to classifier training and model evaluation. The problem has been studied from both unary as well as binary classification perspectives.

The problem of compromised account detection have mostly been studied from a binary classification perspective considering data samples belonging to same user as positive class and that belonging to spam messages or other random users as negative class [4,17,29]. Although this policy has been widely practiced to create a synthesized negative class data, yet it becomes quite difficult to obtain a very-well sampled negative data. Obtaining the negative class in case of problems such as compromised account detection is limitless and it is often difficult to cover all the possible abnormal behaviors. Also, with this limited negative class data, any new abnormal behavior not previously learned in the training phase would be incorrectly detected at run time. Hence, in this research, apart from studying the problem from binary classification perspective it has also been studied from one-class perspective where only the positive data samples have been used for profiling and training of models. Also, efficiency of respective unary and binary machine learning classifiers have been analyzed in order to yield the best classifier to be considered for the automated process.

3.1   Approach Overview

Figure 1 illustrates the profiling based approach followed in the work. Data collected from Twitter has been used for conducting the experiments. But the proposed approach is a generalized one and can be deployed on any social network platform. For each user, a behavioral profile is built that comprise of the meta data information extracted from the historical tweets of that user. Depending upon the number of meta data dimensions (or features) being considered, respective number of models are constructed from which normal behavioral patterns are determined. Continuous and periodic monitoring of the profile need to be performed to determine whether the user is still adhering to the modeled profile or not. In case of a deviation from the baseline profile, the trained models could mark the deviation to be anomalous and hence compromised. In case compromisation of account is indicated, the user can be accordingly notified to take the required action. For a new unknown tweet sample, the objective is to determine whether the activity under consideration has been performed by the concerned user Ui or not. For that to happen, the process can be divided into three steps.

- Firstly, the relevant features are extracted and profiled from the existing tweet samples of a user. A baseline profile is created for the user's account which helps to define a normal usage pattern followed by a user. Each user must have enough data samples so as to have sufficient labeled training data.

- Similar features extracted from the unknown sample are compared with the behavioral profile to compute anomaly score. Accordingly, the unknown tweet sample is assigned the class having the most matched feature values. For binary classification, the tweet sample is assigned that class to which its anomalous score is adhering to the matching criteria. Similarly, in unary classification, if the feature value does not match the positive class, it is counted as anomalous and the account is detected to be probably compromised. Unlike other domains, here the classifiers are trained and tested independently for each respective user because of the consistency maintained by a user on his/her own behavioral profile. This approach is scalable to a large number of users since each user is independently trained and tested.
- For incorporating the dynamics and changing behavior of users, an incremental learning process is adopted by continuously updating the training data with new patterns.

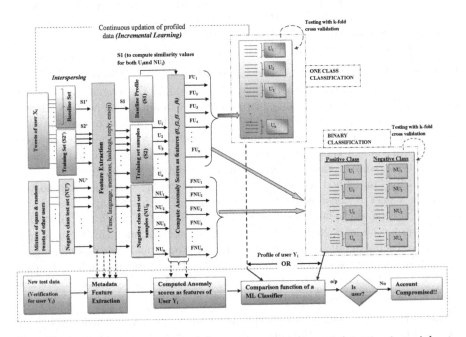

Fig. 1. Proposed MetaCom approach for compromised account detection in social networks

3.2   Construction of Behavioral Profile

This work involves the use of various meta data features for profiling the behavioral pattern of a user. Experiments have been performed on Twitter platform by extracting the meta data information associated with each tweet. To construct the behavioral profile of a user following procedure is adopted.

Firstly, for a user U, a set of tweets is collected from Twitter and sorted in chronological order. These tweets help in building the respective behavioral profile of that user. A set of meta data characteristics such as language, time, emoji etc. is extracted and using these characteristics, respective models are trained. Around 2000 tweets of each user were used to construct baseline profile which help profile the versatility in behaviour of a user and variation in behavior patterns over time. Once a behavioral profile is constructed, the incoming posts are compared to it to assess the extent to which new incoming message complies to the learned behavioral profile. For the same, anomalous scores are computed for each post by extracting and then comparing the feature values to the respective behavioral profile models. The anomalous scores corresponding to each feature act as numerical feature vectors to train the machine learning models.

3.3   Feature Extraction and Modeling

Eight features namely, language, time of the day, day of the week, mentions, hashtags, replyto, emoji and URL has been used for behavioral profiling and anomaly score calculation. A Script in Python is written to collect the tweets and the basic meta data characteristics corresponding to each tweet.

Language: Most of the social networks provide flexibility to users to write their messages in any language of their choice. But based on the language proficiency users usually adhere to just one or two languages to author their posts. Hence, for such profiles where users abide by only a few languages to communicate, a sudden change in language may indicate something anomalous. In the meta data information of each tweet, Twitter assigns a machine detected language to each tweet and in case it seems difficult to detect any language the label 'und' stating 'undefined' is assigned. Behavioral profile corresponding to this language aspect is constructed using the count of each language used. As an example, format of the language profile is $\{English : 200, French : 64, German : 54, Chinese : 2, und : 20\}$. Though, machine detected language is not always correct and reliable but it is quite time intensive to manually check the assigned labels hence, language identifier of those tweets who are assigned to less than 2% language count is changed to 'und'. For the training set tweets, the anomalous score is calculated using Algorithm 1.

Time of the Day: Time profile particularly hour of the day captures how often a user tweets at a particular time of the day. Foundation stone for this assumption has been the concept of circadian typology or chronotype behavior [1] defining the difference in individual's personality. Some users are seen to be active on the social network platforms at certain periods during the day (e.g. early morning, midday, late night etc.). During profiling, if regularity of time in tweets is found, presence of a tweet in inactive hours indicates the anomalous behavior. To build the time profiles, discretization of time is performed. 24 time intervals with one hour length (00–01 h, 00–02 h, ..., 12–13 h, 13–14 h, ..., 23–00 h) are created and accordingly the timestamp of tweet is assigned to one of those intervals. Because of the discretization of time, the adjustment step followed by Eagle et al. [9] is

also implemented in this work. For each time period t, values corresponding to two adjacent hours are also considered i.e. actual value at t^{th} hour amounts to the average of count of tweets during t^{th} hour, $(t-1)^{th}$ hour and $(t+1)^{th}$ hour.

Day of the Week: This attribute helps in capturing the per day behavior of each user as some users may be more active on particular days based on their working schedule (e.g. weekdays or weekends). As in *time of the day* attribute, here also, discretization has been performed into seven days each with duration of one day ($Monday, Tuesday, ..., Sunday$). Count of tweets in each day help analyze the tweeting behavior of the user during particular days. For train and test set tweets, anomalous score value is again calculated using Algorithm 1.

Algorithm 1. Anomalous Score for an entity (Language, Time of the day, Day of the week // *Mandatory Entities*)

Input: *BSet_Entity : Entity values in the Baseline Set, BSet_Entity_count : List with the count of each entity value */*
Output: An$_{score}$*: Anomalous score produced for the corresponding entity in a tweet*
/* n is the number of respective entity values in the tweet. Same procedure is adopted for each of the entities such as Language, Time of the Day, Day of the week. Hence, a generalized algorithm is proposed /*

```
 1: if Entity[i] in BSet_Entity then
 2:     ind = index(BSet_Entity, Entity[i])
 3:     count = BSet_Entity_count[ind]
 4:     if count ≥ Entity_avg then
 5:         an_Entity = 0
 6:     else
 7:         Entity_diff = Entity_avg - count
 8:         an_Entity = Entity_diff / (Entity_avg + Entity_diff)
 9:     end if
10: else
11:     an_Entity = 1
12: end if
13: An_score = an_Entity
        Fetch the index of an item

14: Function fetch_index(BSet_Entity, ent)
15: for j = 1 to n do
16:     if ent == BSet_Entity[j] then
17:         return j
18:     end if
19: end for
20: EndFunction
```

$$7: \quad Entity_{diff} = Entity_{avg} - count$$
$$8: \quad an_{Entity} = \frac{Entity_{diff}}{Entity_{avg} + Entity_{diff}}$$

From Algorithm 1 following points can be devised:

1. For each respective entity if the entity value is already present in the baseline set, then the count value of that entity is compared to the average of all the value counts. If the count for that entity is greater than the average count, then the anomalous score of 0 is assigned as sufficient number of benign tweets with this entity value are already present and hence, this should not be considered anomalous as per the adopted policy.
2. On the other hand, if the value is less than the average value count, then the tweet is categorized as somewhat anomalous and an anomalous score equivalent to the proportion of difference from average value is returned.

3. If the entity value is not present in the baseline set then an anomalous score of 1 is returned as no tweet with the respective entity value is found in the baseline set.

Mentions: Mentions act as a mechanism to directly interact with other users in Twitter. In order to do so, users have to directly place other user's username preceeded by @ symbol to mention the user in the message. Using this process, over time a user creates a personalized interaction history with others which helps to profile the interaction behavior of the user by keeping track of the users mostly in touch with the current user. Again the frequency count of each mention by the user is profiled and checked against anomalous scores for upcoming tweets. Algorithm 2 is adopted to compute anomalous score for 'mention' entity.

Hashtags: Hashtag (#) is a popular entity in Twitter used to define the topic of the tweet. It is hypothesized that users typically have a predefined range of topics that are frequently used in their tweets for example, favourite TV shows, models, sports, politics. For such users who usually tweet about some particular topics and suddenly when their profile is seen floating a tweet about an unrelated topic, it should be considered anomalous. While constructing the hashtag behavioral profile all the hashtags used in the baseline set are mapped with their frequency counts. Using this behavioral profile, anomalous score (An_{score}) for each individual tweet is calculated as proposed in Algorithm 2.

replyTo: This entity helps to keep track of the users whose tweets the user in question usually replies to. This entity originates from the functionality of Twitter to let users reply to the tweets of other users. As in mentions, it is seen that users mostly follow a defined set of users or groups (at least belonging to a fixed number of subjects). Therefore, profiling of this entity seems viable. A sudden deviation in behavior of replyTo attribute may mark the point of anomaly and thus the compromisation of account.

Emoji: Emoji is often considered a symbolic language that let users express themselves in a creative and fun-loving way. As per a report from Swyft Media[1], around 6 billion emoticons are being used in a day on different platforms by people all over the world. Moreover, it has been observed that with the increase in emoji usage, the counterpart slangs such as LOL, rofl etc. are loosing their hold as people are preferring to opt for the former. Type of emoji used by an individual is influenced by the basic human nature and the type of communication a person is usually involved in. Also, other factors such as location, gender, age and social status help in judging the choice of emojis used by an individual.

For emoji profile construction, apart from storing the presence of emoji in a tweet, the status of each emoji used by an individual followed by its count is stored. Further, procedure defined in Algorithm 2 is adopted to calculate anomalous score for each upcoming tweet.

URL: Twitter provides the option to include URLs in the tweets. When meta data is fetched from twitter, information regarding the presence of a URL is

---

[1]https://thenextweb.com/insider/2015/06/23/the-psychology-of-emojis.

indicated by tagging the attribute "URL-present" as true or false defining the presence or absence of URL in the tweet respectively. This presence of tweet is supplemented with the domain counts as features. Python "urlparse" package is used to extract the domain name from URLs used in the tweet. Even the actual domains of the tinyUrls are fetched and stored. Anomalous score for the forthcoming tweets is fetched in the similar fashion as defined in Algorithm 2.

Algorithm 2. Anomalous Score for an entity (Hashtags, Mentions, replyTo, Emoji, URL // *Optional Entities*)

Input: *BSet_Entity : Entity values in the Baseline Set, BSet_Entity_count : List with the count of each entity value */*
Output: An$_{score}$: *Anomalous score produced for the corresponding entity in a tweet*
Initialization :
/* *n is the number of respective entity values in the tweet. Same procedure is adopted for each of the entities such as Mentions, Hashtags, replyTo, Emoji and URL. Hence, a generalized algorithm is proposed /**

1: **for** $i = 1$ to n **do**
2:   **if** Entity[i] == TRUE **then**
3:     **if** Entity[i] in BSet_Entity **then**
4:       ind = index(BSet_Entity, Entity[i])
5:       count = BSet_Entity_count[ind]
6:       $a1 = \frac{count(Entity(True))}{count(Entity(True)) + count(Entity(False))}$
7:       $a2 = 1 - \frac{count}{sum(BSet\_Entity) - count(Entity(False))}$
8:       an$_{Entity}$ = a1 * a2
9:       Entity$_{sum}$.append(an$_{Entity}$)
10:     **else**
11:       an$_{Entity}$ = 1
12:       Entity$_{sum}$.append(an$_{Entity}$)
13:     **end if**
14:   **else**
15:     an$_{Entity}$ = $\frac{count(Entity(False))}{count(Entity(True) + count(Entity(False))}$
16:     Entity$_{sum}$.append(an$_{Entity}$)
17:   **end if**
18: **end for**
19: An$_{score}$ = sum(Entity$_{sum}$)

  Fetch the index of an item

20: **Function** *fetch_index*(BSet_Entity, ent)
21: **for** $j = 1$ to n **do**
22:   **if** ent == BSet_Entity[j] **then**
23:     **return** j
24:   **end if**
25: **end for**
26: EndFunction

Overall from Algorithm 2, following points can be devised:

1. For each respective entity if the tweet does not contain the particular entity, then the probability of tweet not containing that particular entity is used to compute the anomalous score.
2. In case the entity is present, then the value is compared to the values stored in the baseline set. If an entity value is not present in the profile of a user, then anomalous score of 1 is assigned as the profile has not encountered such value before and hence, is suspected to be suspicious.
3. But if the value is already present in the profile, probability of the presence of entity in a tweet along with the score accumulated by that entity value is used to compute the anomaly score.

Anomalous scores are further used as numerical features which are input to the respective classifiers for further processing. RobustScaler is used to normalize the features as it normalizes the features in the range of 0 to 1 by replacing the feature value using $val_{new} = (val - Q_1(val))/(Q_3(val) - Q_1(val))$ where Q defines the interquartile range. This normalization is preferred to reduce the effect of outliers which may be present in the data.

4   Experimental Results and Discussion

Experiments have been performed on the Twitter data. Twitter API is used to extract tweets and the corresponding meta data information from Twitter. Data of some 2000 users with a maximum of 4000 tweets have been used for experiments. For the experimental results the problem has been studied from both binary as well as unary classification perspectives. For binary classification, efficiency of k-Nearest Neighbor (kNN), Random Forest, Gradient Boosted, SVM and MLP classifiers have been analyzed. Likewise, in unary classification, popular one-class classifiers namely, Local Outlier Factor (LOF), Isolation Forest (IF) and One-class SVM have been studied and analyzed. Adopted parameter tuning of these classifiers is as per Table 1. Use of these classifiers have been encouraged in the literature [3, 4, 6, 15, 17, 18, 21, 24, 26] but researchers have mostly deployed these classifiers to handle the said problem using textual features.

4.1   Experimental Approach

In this research work, efficiency of various meta data features and hence these unary and binary classifiers have been studied to detect compromised accounts in online social networks. Experiments have been performed on a 64-bit Windows Operating system with 4.2 GHz i7 processor and 32 GB RAM. All the classifiers are parameter tuned varying the hyper parameters to look for the best combination of parameters for optimal efficiency. All the performed experiments are coded in Python and a machine learning toolkit named scikit-learn [25] is used for the classification tasks. As a common machine learning classification policy, collected data is split into 60% training and 40% testing parts. A Shuffle-Split approach with 5-fold cross-validation is performed on training data (split into 60-40 train and validation set). Every classifier is parameter-tuned and trained with the feature vector obtained from anomalous scores generated from meta data. Once a classifier is trained, test data is fed to each parameter-tuned classifier to validate the model and obtain the evaluation metrics. Performance of the said classifiers have been analysed on the basis of various evaluation metrics and is shown in Tables 2 and 3.

Table 1. Parameter settings of different classifiers

Classifier	Varied Parameters and their chosen values (Notations specific to scikit-learn (Python))
Binary classifiers	
K-Nearest Neighbor	– *n_neighbors:* [3, 5, 7, 9]
Random Forest	– *n_estimators:* [10, 100] – *max_features:* ['auto'] – *max-depth:* [10, 20, 30]
Gradient Boosted	– *n_estimators:* [100, 200] – *max_depth:* [1, 2, 3] – *learning_rate:* [0.01, 0.1]
SVM	– *C:* [0.1, 1, 10, 100] – *Kernel:* ['linear', 'poly','rbf', 'sigmoid' – *gamma:* [0.1, 1]
Multi Layer Preceptron (Neural Networks)	– *activation function:* ['tanh', 'relu'] – *alpha:* [0.05, 0.1, 0.5] – *hidden_layer_sizes:* [(50, 50), (50, 100, 50), (100)]
Unary classifiers	
Local Outlier Factor (LOF)	– *n_neighbors:* [3, 5, 7, 9]
Isolation Forest (IF)	– *n_estimators:* [10, 100]
OCC-SVM	– *Gamma:* [0.1, 1, 5, 10], – *nu:* [0.1, 0.2, 0.3], – *kernel:* ['linear', 'poly', 'rbf', 'sigmoid']

4.2 Results

Major aim of this research work is to analyze the efficacy of different meta data features and machine learning classifiers to detect compromised accounts in social networks. The task is to asses which features and algorithm as well as what set of fine tuned parameters help achieve better results. Depending on the availability and authenticity of the negative class data, the problem has been studied from both binary and unary class classification perspectives. For the binary class classification, meta data features extracted from the data samples of the same user are taken as positive samples and the mixture of meta data features from data of other random users and some random spam tweets are considered as negative class. Using such meta data models, anomalous scores calculated for each corresponding tweet are taken as feature vectors to train the respective classifiers.

Table 2. Performance of different binary classifiers with their optimized parameters

Classifier	Optimized Parameters for the classifier	Best Parameter Setting	Accuracy (%)	Precision (%)	Recall (%)	F-score (%)	FBeta Score (%)	ROC-AUC (%)
kNN(neighbors)	*Neighbors:* [3,5,7,9]	kNN(9)	91.75 ± 3.5	91.39 ± 4.0	90.78 ± 4.8	90.86 ± 4.2	91.12 ± 4.0	96.18 ± 2.1
RF(max_depth, max_features, n_estimators)	*max_depth:* [10,20,30], *max_features:* ['auto'], *n_estimators:* [10,100]	RF(30,'auto',100)	93.28 ± 3.1	92.19 ± 4.4	93.44 ± 4.9	92.66 ± 3.5	92.34 ± 3.8	97.80 ± 1.6
GB(learning_rate, max_depth, n_estimators)	*learning_rate:* [0.01,0.1], *max_depth:* [1,2,3], *n_estimators:* [100,200]	GB(0.1,1,200)	92.31 ± 3.4	91.42 ± 4.4	92.08 ± 6.4	91.54 ± 4.1	91.41 ± 3.8	97.34 ± 1.9
SVM(C, gamma, kernel)	*C:* [0.1,1,10,100], *gamma:* [0.1,1], *kernel:*['linear','poly','rbf','sigmoid']	SVM (10,1,'rbf')	88.67 ± 4.8	87.07 ± 6.1	89.03 ± 9.2	87.60 ± 6.1	87.15 ± 5.4	94.68 ± 3.3
MLP(activation, alpha, hidden_layer_sizes, learning_rate)	*Activation:*['tanh','relu'], *alpha:* [0.05, 0.1, 0.5], *hidden_layer_sizes:*[(50,50), (50,100,50),(100,)]	MLP('tanh',0.05, (50,100,50))	88.10 ± 4.6	86.98 ± 5.5	87.55 ± 9.2	86.75 ± 6.5	86.71 ± 5.3	93.89 ± 3.8

Table 3. Performance of different unary classifiers with their optimized parameters

Classifier	Optimized Parameters	Best Parameter Setting	Accuracy (%)	Precision (%)	Recall (%)	F-score (%)	FBetaScore (%)
LOF(n_neighbors)	*Neighbors:* [3,5,7,9]	kNN(9)	65.43 ± 12.67	91.78 ± 15.04	59.39 ± 13.07	70.54 ± 13.95	63.84 ± 13.84
IF(n_estimators)	*n_estimators:* [10,100]	IF(100)	68.37 ± 11.85	89.08 ± 0.2	62.37 ± 12.43	70.99 ± 8.79	67.58 ± 12.60
OCC-SVM(nu, gamma, kernel)	*nu:*[0.1,0.2,0.3], *gamma:*[0.1,1,5,10], *kernel:* ['linear','poly','rbf','sigmoid']	OCC-SVM (0.2,5,'rbf')	74.02 ± 9.37	74.86 ± 10.07	73.19 ± 13.71	72.72 ± 9.36	73.47 ± 9.44

From Table 2 it is observed that if the problem is studied as a binary classification problem, every respective fine-tuned classifier attained a good performance (above 86 % F-score) on their respective setting of parameters. However,
it is still evident that Random Forest outperformed other classifiers in terms of
different performance parameters. Also, unlike existing literature, the problem
has also been studied from one-class perspective where data samples belonging
to only the positive class are considered to train the respective classifiers. For
the problems such as detection of compromised accounts identification of both
true positives as well as true negatives is important. Hence, in order to test the
efficiency of the trained models in correctly categorizing both compromised as
well as non-compromised tweets, the negative class data of random users and
spam tweets was brought into practice in the test set. Training set however only
comprise of the positive class data samples, hence the models are only trained
with positive data samples making it a purely unary approach.

From Table 3, it is seen that out of three one-class classifiers, namely, LOF,
IF and OCC-SVM, OCC-SVM outperformed others attaining an average F-
score of 72.72%. Values in Tables 2 and 3 represent the average score of different
evaluation metrics for all the users. In order to demonstrate the performance
of each fine tuned classifier for each respective user, boxplot (Fig. 2) for both
binary and unary classifiers is shown with 25 to 75% interquartile range. For
each classifier most of the values skewed to the higher Fscore range and for some
25% users, Fscore values ranged from lower quartile to the minimum whisker
(which even in worst case i.e. of MLP is around 80%). Only a small amount of
users obtained F-score less than 87% with Random Forest. Similarly, for OCC-
SVM in one-class classification, performance for users is varying ranging between
lower quartile of 58% to upper quartile of 90%.

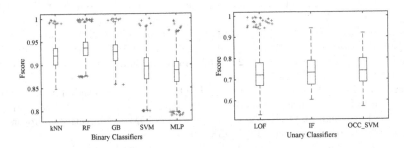

Fig. 2. Box plots depicting variation of performance metrics with different binary and
unary classifiers

5 Conclusion

This work examined the efficiency of various meta data features in detecting the
compromised accounts in social networks. Although Twitter platform has been

chosen to perform the said experiments yet the technique can be deployed on any social network as the considered meta data features are commonly present in almost all popular social networks. Problem has been studied from both unary as well as binary classification perspectives and hence, the efficiency of both one-class as well as two-class classification techniques have been examined. It has been observed that amongst various binary classifiers Random Forest performed most efficiently for the undertaken problem attaining an average F-score of 92.66%. Similarly, amongst three one-class classifiers, OCC-SVM with 'rbf' kernel achieved 72.72% average F-score better than the other classifiers.

Acknowledgment. This publication is an outcome of the R&D work under Visvesvaraya PhD Scheme of Ministry of Electronics & Information Technology, Government of India, being implemented by Digital India Corporation under Grant Number: PhD/MLA/4(61)/2015-16.

References

1. Adan, A., Archer, S.N., Hidalgo, M.P., Di Milia, L., Natale, V., Randler, C.: Circadian typology: a comprehensive review. Chronobiol. Int. **29**(9), 1153–1175 (2012)
2. Adewole, K.S., Anuar, N.B., Kamsin, A., Varathan, K.D., Razak, S.A.: Malicious accounts: dark of the social networks. J. Netw. Comput. Appl. **79**, 41–67 (2017)
3. Al-Ayyoub, M., Jararweh, Y., Rabab'ah, A., Aldwairi, M.: Feature extraction and selection for arabic tweets authorship authentication. J. Ambient Intell. Humaniz. Comput. **8**(3), 383–393 (2017)
4. Barbon, S., Igawa, R.A., Zarpelão, B.B.: Authorship verification applied to detection of compromised accounts on online social networks. Multimed. Tools Appl. **76**(3), 3213–3233 (2017)
5. Brocardo, M.L., Traore, I., Woungang, I.: Authorship verification of e-mail and tweet messages applied for continuous authentication. J. Comput. Syst. Sci. **81**(8), 1429–1440 (2015)
6. Brocardo, M.L., Traore, I., Woungang, I.: Continuous authentication using writing style. In: Obaidat, M.S., Traore, I., Woungang, I. (eds.) Biometric-Based Physical and Cybersecurity Systems, pp. 211–232. Springer, Cham (2019). https://doi.org/10.1007/978-3-319-98734-7_8
7. Brocardo, M.L., Traore, I., Woungang, I., Obaidat, M.S.: Authorship verification using deep belief network systems. Int. J. Commun. Syst. **30**(12), e3259 (2017)
8. Egele, M., Stringhini, G., Kruegel, C., Vigna, G.: COMPA: detecting compromised accounts on social networks. In: NDSS (2013)
9. Egele, M., Stringhini, G., Kruegel, C., Vigna, G.: Towards detecting compromised accounts on social networks. IEEE Trans. Dependable Secure Comput. **14**(4), 447–460 (2017)
10. Farseev, A., Nie, L., Akbari, M., Chua, T.S.: Harvesting multiple sources for user profile learning: a big data study. In: Proceedings of the 5th ACM on International Conference on Multimedia Retrieval, pp. 235–242. ACM (2015)
11. Fire, M., Kagan, D., Elyashar, A., Elovici, Y.: Friend or Foe? Fake profile identification in online social networks. Soc. Netw. Anal. Min. **4**(1), 194 (2014)
12. Gao, H., Hu, J., Wilson, C., Li, Z., Chen, Y., Zhao, B.Y.: Detecting and characterizing social spam campaigns. In: Proceedings of the 10th ACM SIGCOMM Conference on Internet Measurement, pp. 35–47. ACM (2010)

13. Hu, X., Li, B., Zhang, Y., Zhou, C., Ma, H.: Detecting compromised email accounts from the perspective of graph topology. In: Proceedings of the 11th International Conference on Future Internet Technologies, pp. 76–82. ACM (2016)
14. Igawa, R.A., Almeida, A., Zarpelão, B., Barbon Jr., S.: Recognition on online social network by user's writing style. iSys-Revista Brasileira de Sistemas de Informação 8(3), 64–85 (2016)
15. Jankowska, M., Keselj, V., Milios, E.: Proximity based one-class classification with common n-gram dissimilarity for authorship verification task. In: CLEF 2013 Evaluation Labs and Workshop-Working Notes Papers, pp. 23–26 (2013)
16. Johansson, F., Kaati, L., Shrestha, A.: Time profiles for identifying users in online environments. In: 2014 IEEE Joint Intelligence and Security Informatics Conference, pp. 83–90. IEEE (2014)
17. Kaur, R., Singh, S., Kumar, H.: AuthCom: authorship verification and compromised account detection in online social networks using ahp-topsis embedded profiling based technique. Expert Syst. Appl. 113, 397–414 (2018)
18. Koppel, M., Schler, J.: Authorship verification as a one-class classification problem. In: Proceedings of the Twenty-First International Conference on Machine Learning, p. 62. ACM (2004)
19. Laleh, N., Carminati, B., Ferrari, E.: Anomalous change detection in time-evolving OSNs. In: 2016 Mediterranean Ad Hoc Networking Workshop (Med-Hoc-Net), pp. 1–8. IEEE (2016)
20. Lee, S., Kim, J.: Warningbird: a near real-time detection system for suspicious urls in twitter stream. IEEE Trans. Dependable Secure Comput. 10(3), 183–195 (2013)
21. Li, J.S., Chen, L.C., Monaco, J.V., Singh, P., Tappert, C.C.: A comparison of classifiers and features for authorship authentication of social networking messages. Concurr. Comput. Pract. Exp. 29(14), 1–15 (2016)
22. Nauta, M.: Detecting hacked twitter accounts by examining behavioural change using twitter metadata (2016)
23. Nauta, M., Habib, M., van Keulen, M.: Detecting hacked twitter accounts based on behavioural change. In: Proceedings of the 13th International Conference on Web Information Systems and Technologies, pp. 19–31 (2017)
24. Neal, T., Sundararajan, K., Woodard, D.: Exploiting linguistic style as a cognitive biometric for continuous verification. In: 2018 International Conference on Biometrics (ICB), pp. 270–276. IEEE (2018)
25. Pedregosa, F., et al.: Scikit-learn: machine learning in Python. J. Mach. Learn. Res. 12, 2825–2830 (2011)
26. Peng, J., Choo, K.K.R., Ashman, H.: Bit-level n-gram based forensic authorship analysis on social media: Identifying individuals from linguistic profiles. J. Netw. Comput. Appl. 70, 171–182 (2016)
27. Ruan, X., Wu, Z., Wang, H., Jajodia, S.: Profiling online social behaviors for compromised account detection. IEEE Trans. Inf. Forensics Secur. 11(1), 176–187 (2015)
28. Sahoo, S.R., Gupta, B.B.: Classification of various attacks and their defence mechanism in online social networks: a survey. Enterp. Inf. Syst. 13(6), 832–864 (2019)
29. Tröng, D., Johansson, F., Rosell, M.: Evaluating algorithms for detection of compromised social media user accounts. In: 2015 Second European Network Intelligence Conference (ENIC), pp. 75–82. IEEE (2015)

30. VanDam, C., Tang, J., Tan, P.N.: Understanding compromised accounts on twitter. In: Proceedings of the International Conference on Web Intelligence, pp. 737–744. ACM (2017)
31. Viswanath, B., et al.: Towards detecting anomalous user behavior in online social networks. In: 23rd USENIX Security Symposium (USENIX Security 2014), pp. 223–238 (2014)

Ameliorating Depression Anxiety and Stress Caused by Social Media Among Young Women: Procedural and Technological Responses

Honour Carmichael, Gabrielle Peko, Khushbu Tilvawala,
Johnny Chan, and David Sundaram[(✉)]

Department of Information Systems and Operations Management,
University of Auckland, Auckland, New Zealand
hcar959@aucklanduni.ac.nz, {g.peko, k.tilvawala,
jh.chan, d.sundaram}@auckland.ac.nz

Abstract. Young women aged between 16–24 are often severely affected by social media, causing them to become depressed, anxious and stressful. The current procedural and technological responses to depression, anxiety and stress (DAS) are not directly targeting young women, even though one in four experience symptoms of common mental health conditions. Furthermore, there are limited studies on DAS among women of this age group. Following the design science research methodology, conceptual and system artefacts are created and explored to improve current responses to ameliorate DAS caused by social media for young women in this paper. We propose the Happiness App as a solution that caters to young women negatively affected by social media. These artefacts are evaluated using the V-model framework. Our findings reinforce that social media is an environment which could create DAS among young women. Difficulties emerge when individuals attempt to break their habit of using social media. Individuals, organizations and governments are all accountable in ensuring that social media is a safe space for all.

Keywords: Stress · Anxiety · Depression · Social media · Young women · Social network · Happiness App

1 Introduction

Social media has become a paramount part of an individual's daily life, making it difficult to live without internet access or a smartphone. Constant connectivity to others increases the need for gratification and affirmation, which may cause depression, anxiety and stress (DAS) among people. Individuals experience varying levels of DAS, making it difficult for a person to measure and recognise the related symptoms. Young women aged 16–24 are highly vulnerable to DAS. The motivation of this paper is to ameliorate DAS among young women caused by social media. By understanding their habits and actions with social media, this paper aims to provide frameworks to improve the procedural and technological responses towards DAS. It first reviews the existing

R. Doss et al. (Eds.): FNSS 2019, CCIS 1113, pp. 81–96, 2019.
https://doi.org/10.1007/978-3-030-34353-8_6

literature. Following that, conceptual and system artefacts are created, and a proof of concept is designed. Lastly, the consequences of DAS and social media for young women are discussed.

2 Literature Review

This section aims to review relevant literature on DAS, young women and social media. It focuses on how DAS affects young women, how social media could cause DAS in young women, and the current procedural and technological responses. A critical review of these sections leads to the discovery of the problems, issue and requirements (Fig. 1).

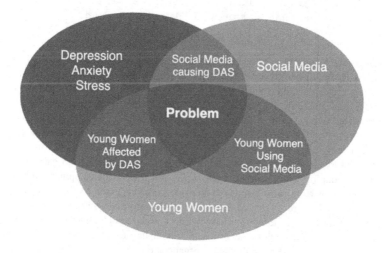

Fig. 1. Research problem

2.1 Depression, Anxiety and Stress (DAS)

Depression is when a person feels down over weeks or months at a time. Symptoms of depression include hopelessness, loss of enjoyment and interest, negative thinking and sleep problems. Depression can affect regions of the brain through inflammation and oxygen reduction. Self-harm and suicide can be a result of depression. There are different types of depression including Major Depressive Disorder, Persistent Depressive Disorder and Seasonal Affective Disorder. In New Zealand, 1 in 5 women experience depression and 1 in 7 experience depression before they are 24 years old [1].

Anxiety is when a person feels nervous due to worries and fears that may replay in one's mind. Physical symptoms include heart palpitations, stomach cramps, sweating and light-headedness. Anxiety is generally intermittent depending on the person. An anxiety disorder is where chronic or irrational fear prevents a person from performing daily activities. There are several types of anxiety disorders including Generalized Anxiety Disorder (GAD), Obsessive Compulsive Disorder (OCD), Post-Traumatic

Stress Disorder (PTSD) and Panic Disorder. In New Zealand, 6.1% of people have been diagnosed with anxiety disorder at some time in their lives [2].

Stress is a physical, mental and emotional response when the individual perceives that they cannot cope with demands. Physical symptoms include stomach pain, sleeplessness and fatigue. Mentally, symptoms include restlessness, irritability and anxiousness. Negative thoughts, beliefs and attitudes also cause individuals to develop stress coping mechanisms such as smoking, overeating and drinking. In the long term, stress affects health, productivity and well-being, causing high blood pressure, heart disease, obesity and diabetes [3]. Stress management is essential as the level of stress experienced varies through one's life.

DAS are commonly experienced at the same time. In New Zealand, 1 in 5 people experience depression and anxiety at the same time [1]. According to the Health Loss in New Zealand study, anxiety and depressive disorders are the second leading cause of health loss, accounting for 5.3% of all health loss [4]. DAS affects more than one aspect of a person, as physical, mental, social and spiritual aspects work together. Stress can also be a symptom of anxiety and depression, increasing negative emotions.

2.2   DAS Among Young Women

Studies have found that women experience more symptoms of DAS than men. Statistics collected by the National Health Service England show that one in four women aged 16–24 reported symptoms of common mental health conditions [5]. Between the ages of 16–24, significant life changes occur as young women transition into adulthood. A 5-year longitudinal study of 17 or 18-year-old women concluded that 47% of the women had one or more episodes of major depression [6]. As many of those aged 16–24 are high school and university students, it is important to consider the pressures of studying as a contributor to DAS. Undergraduate students who reported higher levels of stress were less likely to make healthy food choices such as consuming fruits and vegetables [7]. Anxiety levels were found to be higher in female students [8]. Women suffering from DAS are also attempting to seek help. However, the Adult Psychiatric Morbidity survey revealed that 17.9% of women have been diagnosed with depression at some time in their lives and 7.7% with anxiety disorder and that despite going to their general practitioner for help, only 20% of young women received treatment [9]. These results show that DAS is prevalent among young women and their need for help is not being met.

2.3   DAS Among Young Women Caused by Social Media

The emergence of Web 2.0 in the early 2000s and web-based technologies have enabled social media to become a dominant force in society today. Platforms such as Facebook, Instagram, WhatsApp, YouTube and WeChat are common places for Millennials and Generation Z to like and comment on each other's photos, videos and posts. As many as 90% of young adults in the US use social media and visit it at least once a day [10]. There are mixed findings as to whether social media causes DAS. On

one hand, some studies found that not only are there no negative correlations between Facebook activity and depression for females with high levels of neuroticism, but high levels of Facebook activity are associated with lower levels of depression [11]. On the other end, "Facebook Depression" has been a term popularized by the American Academy of Pediatrics (AAP) to describe teens spending a significant amount of time on social media and showing symptoms of depression [12]. Fear of missing out (FOMO) is the use of social media in attempt to avoid feeling out of the loop [13]. It is common for young women to feel FOMO as constant access to social media websites through smartphones makes it easy to miss a trend and be out of the loop if inactive online for even a few hours. FOMO manifests causing anxiety, stress and irritability from social media [14]. Another feature of constant interaction on social media is the creation of a "selfie generation" where young adults feel under pressure over their body image. This pressure about body image and need for constant interaction and approval from others has a significant impact on one's life.

In some cases, social media could lead to irreversible outcomes. In May 2019 a 16 year old girl in Malaysia killed herself after using Instagram's voting poll, asking people to help her decide whether to live or die [15]. This is an extreme example of young women influenced by social media. Cyber bullying and cyber harassment are also common. Anonymity and detachment from real world consequences of one's actions has made it easier for cyberbullies to negatively impact young women's lives via social media. Hashtags on Instagram with male and female name abbreviations for mental health issues create places for like-minded depressed, anxious and suicidal people to negatively encourage each other [16]. Generally, social media companies do not protected users from cyberbullying and, as a result, the depression and anxiety could lead towards death.

2.4   Current Procedural Responses

There are many different procedural responses towards DAS. Personal realization of stress, nervousness or unhappiness in one's life and desire to change, or family members and friends pointing out the symptoms of DAS to young women are essential in seeking help. Realization of symptoms of DAS may make young women seek professional help. For most people, this requires speaking to their general practitioner (GP). As there are no physical tests for anxiety and depression, the GP will ask questions to diagnose and determine the appropriate type of treatment. A blood sample may be taken to rule out other conditions with similar symptoms such as hyperthyroidism. Many of the procedural responses listed below are for anxiety and depression rather than stress. This is because stress is seen as a symptom of anxiety and in many ways less serious than anxiety and depression. The Mental Health Foundation discovered that 137,346 people in New Zealand used mental health and addiction services in 2011 [1]. Most clients were seen face-to-face and district health boards were the largest providers of mental health services [1]. There are many scales used to help measuring DAS. Depression Anxiety Stress Scales is a 42-item questionnaire which participants fill out showing what levels of what they feel [17]. For stress, there is the

Perceived Stress Scale that is used to measure the thoughts and feelings of a person over the last month with 10 questions. For anxiety, there is the General Anxiety Disorder 7-item scale [18]. One of the most used anxiety scales is the Hamilton Anxiety Rating Scale (HAM-A) [19]. Other depression scales include the Kessler 10 and the PHQ-9 Patient Health Questionnaire [20]. If diagnosed with mild depression, watchful waiting may be applied to see if the symptoms get better by itself. This means waiting another 2 weeks and seeing the GP again. This can be frustrating for individuals as it may have taken a lot of courage to see the doctor in the first place.

Medication: Antidepressant medication is not only prescribed to people experiencing depression, it is also used to help stress and anxiety. Antidepressants are safe and not addictive; however, some people do experience side effects. Common side effects include nausea, headaches, anxiety, sweating, dizziness, weight gain and sexual difficulties [21]. Therefore, it requires discussion and adjustment of medication with doctors. The time frame between a person starting to take antidepressant medication and responding to treatment can take several weeks. Those who have experienced suicidal thoughts, therefore, need to be closely monitored as risk of suicidal behaviour may increase. Antidepressants were prescribed to 427,900 patients in NZ in the year to 30 June 2013, representing more than a 20% increase in the last five years [1]. Benzodiazepines or sleeping pills are prescribed to help people relax and reduce stress in the short term. If taken for a long time the medication can be addictive and affect coordination. It should not be used as the first or only treatment and should be part of a broad treatment plan [22]. Selective Serotonin Reuptake Inhibitors (SSRIs) help increase the level of serotonin thought to be a "good mood" chemical. Side effects include nausea, headaches and dry mouth, however they, usually improve over time. They aren't suitable for people under 18 years of age as research shows the risk of self-harm and suicide may be increased if they're under 18 [22].

Therapies: *Cognitive Behavioural Therapy* (CBT) is a structured psychological treatment that recognizes that the way we think and act could affect the way that we feel [23]. CBT involves working with professionals to identify thought and behaviour patterns that either make the individual more likely to become anxious or stop them from recovering. Changes can then be made to reduce anxiety by replacing the habits and improving coping skills. CBT focuses on problem solving and teaching people how to let go of their worries. *Acceptance and Commitment Therapy* (ACT) helps individuals to accept difficulties in their lives. It is a type of mindfulness-based therapy that encourages acceptance, choosing a valid direction and taking action. It encourages activities that are rewarding, pleasant or giving sense of satisfaction. It also helps individuals to cope with fearful situations rather than avoiding or escaping them [24]. *Interpersonal Therapy* (IPT) helps people focus on how they are feeling and behaving in their relationships [23]. This helps the individual to understand what they are finding difficult and how their DAS is affecting them, and the people close to them. IPT is ideal for moderate to severe depression. *Counselling* is where trained professionals help one to talk about the problems he or she faces personally, socially and psychologically. They also help the individual find new ways to think and support them to experience positive change.

2.5 Current Technological Responses

In the following paragraphs we explore e-therapies, websites, and helplines as examples of technological responses.

E-therapies are computer-aided therapies that are as effective as face-to-face therapy for mild to moderate anxiety and depression. Most e-therapies follow the same principles as CBT and teach users to identify and change patterns of thinking and behaviour that might prevent them from overcoming anxiety. As these therapies are online, they do not require travelling, which maintains the user's privacy about their condition, and they are overall less expensive to deliver. Studies have used a range of technological interventions including desktop computers, automated telephone guided therapies, internet therapies and virtual reality [24]. The most readily available form of therapy for anyone to access is mental health apps. SPARX is an e-therapy tool for ages 12–19 that teaches the skills to help combat depression and anxiety. Using CBT techniques, it allows the user to play games using an avatar at their own pace. Results from a clinical study show significant reduction in depression, anxiety and hopelessness making it an effective self-help resource [25]. VRGET is used to treat phobias, generalized anxiety and panic disorders using virtual reality to simulate situations where the phobia or anxiety would generally occur to help the user to face the situation [26].

Websites and Helplines about DAS are numerous. The mental health foundation website explains DAS and provides useful websites and apps for each of these issues. Depression.org.nz and Anxiety.org.nz both also provide information and quizzes to help assess the levels of depression and anxiety a person has. The Low Down website is another one which helps people with anxiety and depression with a 24/7 helpline through phone, email or webchat. There are many specialist helplines in New Zealand including lifeline, suicide crisis helpline, depression helpline and Samaritans [2].

2.6 Problems, Issues and Requirements

In this section, we summarize the problems and issues found in the literature review and the requirements to overcome these issues.

Problems and Issues: Although there are many different methods from the current procedural responses, not all of them could cater to the different levels of DAS. While the scales of DAS is useful, a person with milder forms of DAS may feel that they do not need to seek help. The "wait-and-see" procedural method is potentially life-threatening especially for those who may be depressed and suicidal without being watched closely. Medication could take time for individuals to adjust and they could have negative side effects, requiring people to trial on different medications. The current technological responses show that there are many e-therapy tools. However many of the mindfulness apps which are easily accessible have not been tested, and therefore it is difficult to know their general effectiveness. Furthermore, some apps cost money and create a barrier of entry. Virtual reality therapy could be effective but they require one to go to a specialist as the software and headset equipment is not cheap nor available for use at home. Social media platforms are addictive and it is clear that they

could be the cause of DAS. Social media companies themselves are only becoming aware of this situation recently. Instagram now puts out warnings for those who search about suicide or depression, but it is one of the few actively doing that. There are apps and extensions that exist to protect children against cyberbullying and suicide. But they are not designed for young women who are more vulnerable to DAS than other demographics.

Requirements: There is a void in the literature on DAS among young women under the influence of social media. A combination of personalized treatments is required for young women aged 16–24 so that they are specifically empowered to combat aspects of social media platforms that make them stressed, anxious and depressed. This tool should be clinically tested and easily accessed from anywhere. By catering to these requirements, young women will be better equipped to ameliorate DAS.

The literature review highlights how depression, anxiety and stress affect young women, one of the most affected demographics. Their interactions with social media create negativity and, in some cases, lead to suicide. Medication and therapy are effective procedural responses towards DAS. Young women who have mild or severe depression and/or anxiety need professional help. But those who experience mild stress may not seek help. These women in particular need effective resources available to them to ameliorate stress and anxiety. Websites and E-therapy provide online and technological solutions, making it easier for them to access help at home.

3 Research Artefacts

Following the design science research methodology [27], designing an app based on the requirements from the previous section for young women to ameliorate DAS is deemed appropriate. This could ensure and enforce positive behaviours towards social media, and enable users to understand and prevent bad habits. This would also educate young women about DAS in order for them to prevent becoming depressed, anxious and stressed, as well as helping other young women. In the following sections, we propose several types of research artefacts, including conceptual and system oriented artefacts such as the Happiness App (Fig. 8), to illustrate different aspects of the proposed solution.

3.1 Conceptual Artefacts

The interrelations among depression, anxiety and stress, and their causal interconnects are illustrated in Fig. 2. Stress and anxiety are both symptoms of depression. Anxiety and depression commonly occur at the same time and stress is largely a symptom of anxiety. Charles Duhigg's Habit loop [28] explains the neurological patterns of one's habit. The cycle starts with a cue that determines which routine to follow. The routine may be mental, emotional or physical, creating a reward that makes the cycle rememberable for the future. For young women who are looking for ways to ameliorate depression, anxiety and stress in their lives, the key in doing so is to change the

addictive habit by replacing the routine. In this case, the routine will be posting or consuming information from social media.

A typical scenario for young women using social media begins with stress [29]. From posting on social media, individuals expect gratification from their peers. The difference between the expected level and the actual level of approval and gratification causes the symptoms of stress to occur. Symptoms include nervousness from waiting for people to like ones post leading to worries about one's self-image and pessimism over their post. Stress causes the individual to take another selfie to satisfy their craving for gratification.

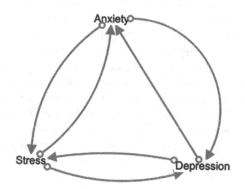

Fig. 2. The interrelation of Depression, Anxiety and Stress

Anxiety is also created from build-up of stress and worrying of self-image. Fear and panic create the urge to post on social media to reduce anxiety. Managing anxiety starts when a person realizes their anxiousness, and they could seek help by using the Happiness App before posting on social media (i.e. replacing the routine).

Depression is an extension of stress and anxiety with emotional instability. The manifestation of negative thoughts causes the individual to take a selfie and post on social media for gratification. The realization of depressive thoughts could trigger a new path of seeking help, which may lead to positive thoughts about one's self and a reduced need to post online.

When the symptoms of depression, anxiety and stress are present, this creates a vicious gratification seeking cycle. Within each cycle, the symptoms become worse. It is unsustainable and unhealthy for young women to experience them. It is evident from the interrelation of DAS that a solution is needed to change the routine.

3.2 Scenarios

In order to explicate the design of the various research artefacts we explored a few scenarios of how young women may typically react to social media.

> *Meghan is 17 and constantly uses social media. She is unaware of the conse-*
> *quences of being on social media and how addicted she is to the smartphone. By*
> *educating Meghan about their social media usage, we can raise her awareness*
> *if she ever experiences symptoms of DAS.*

> *Sarah is 20 years old and uses social media heavily every day, using Instagram*
> *stories to document her daily life in order to receive affirmation due to a lack of*
> *self-confidence. When she does not receive the amount of likes on her post, she*
> *begins to become depressed. When she becomes depressed, she starts to binge*
> *eat. Sarah is aware of the issue but does not know how to fix it. By using the*
> *Happiness App, which allows Sarah to identify when there is a change to nega-*
> *tive emotions and what causes her to binge eat, she can stop the negative cycle.*

> *Denise is 23 years old and uses Facebook daily to check messages from her*
> *friends. She is not concerned about what she posts and does not care about what*
> *people think about her posts. Denise has noticed her best friend, Alice, is anxious*
> *when Denise doesn't like her photo. Although Denise is not affected by social*
> *media, she wants to understand why Alice is anxious and how to help her. By*
> *creating awareness through the Happiness App and providing a platform for*
> *Denise to understand why Alice is acting anxious, she is able to help Alice.*

The scenarios give us an idea of how young women feel and how each of them responds differently towards social media.

3.3   System Oriented Artefacts

System oriented artefacts are frameworks, architectures, mock-us, and/or implementations that help show how a product can be created and implemented to treat DAS. Firstly, we outline the expected process/workflow of young women's amelioration of DAS through the Happiness App (Fig. 3).

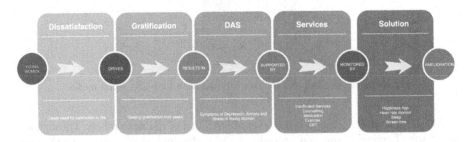

Fig. 3. Process of young women's amelioration of DAS through the Happiness App

The process shows the stages of a young woman going in and out of DAS via social media and the Happiness App. Firstly, a woman enters the stage of dissatisfaction within themselves. This drives them to seek gratification from others through social media. The results of the gratification from peers do not appear to be enough to stop

symptoms of DAS. By using current procedural and technological responses alongside the Happiness App, the symptoms of DAS are ameliorated.

The cue-routine-reward habit loop of DAS among young women is illustrated in Fig. 4. The cue that young women have is unhappiness with themselves; this may include symptoms of DAS such as feeling down. The routine is posting on social media and the reward is the short-term gratification that is felt.

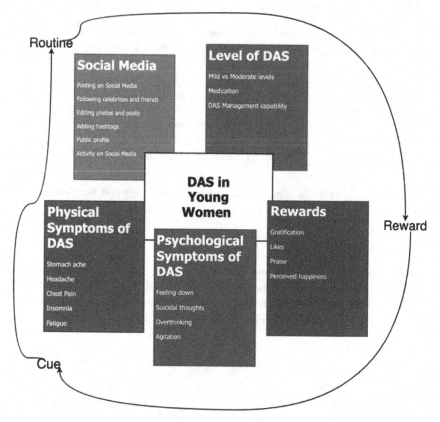

Fig. 4. Areas of DAS that interact with the habit loop and social media

The DAS iceberg in Fig. 5 is the level of DAS that is visible to others. Through the use of the Happiness App and procedural and technological responses to DAS, the DAS iceberg becomes smaller.

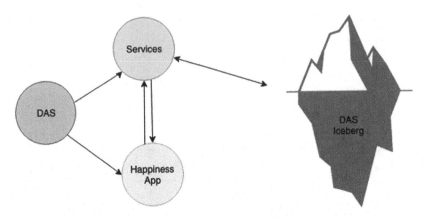

Fig. 5. The Depression Anxiety Stress Iceberg

The stock and flow diagram (Fig. 6) shows the number of people who have a desire to post on social media, which increases with the number of followers. The number of likes expected depends on the expected approval and gratification. The actual number of likes, which is lower than the expected, increases the level of DAS.

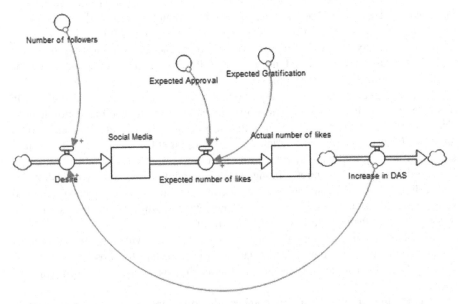

Fig. 6. AS-IS model of DAS

By using the Happiness App (Fig. 7), the level of desire decreases, and therefore decreasing the overall level of DAS. Overall, the system artefacts show how effective implementing the Happiness App will be to decrease the level of DAS in young women.

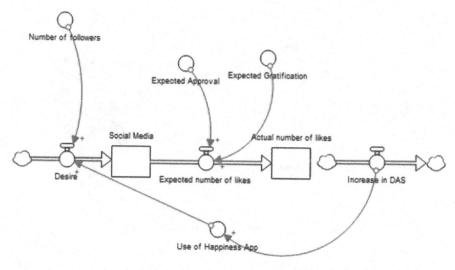

Fig. 7. TO-BE model DAS

3.4   Proof of Concept

To demonstrate the effectiveness and feasibility of ameliorating DAS, we have is designed the Happiness App. The Happiness App requires login and monitoring of social media pages to learn when the user may be experiencing higher levels of depression, anxiety or stress. There are also customizable features allowing the app to cater to the user's personalised needs. The following screenshots demonstrate the features available within the Happiness App.

The home page of the app displays the features that the user has decided to use. In this case, the user has their daily check-in, medication tracking and meditation feature. The medication feature reminds the user to take their medication and allows users to list any side effects or symptoms experienced. At the top right the user can edit the features shown by adding, removing or prioritising features.

The daily check-in requires a rating system and allows users to type their thoughts and feelings so that they are able to measure how social media is affecting them and what techniques to use to combat them. When selecting edit features, the user can go through a list of features, turning them on and off.

The Menstrual tracker not only tracks menstruation and ovulation, but allows users to indicate the emotions they felt that day. This is important for those who are experiencing DAS to help identifying and distinguishing what causes the symptoms of DAS.

Overall, the proof of concept combats the problems, issues and requirements identified from the literature review. The Happiness App provides a customizable platform catered to young women. Other features available on the app include sleep monitoring, heart rate tracker and workout log as exercise significantly improves the symptoms of DAS [30]. This also makes the app flexible and usable on a smartwatch.

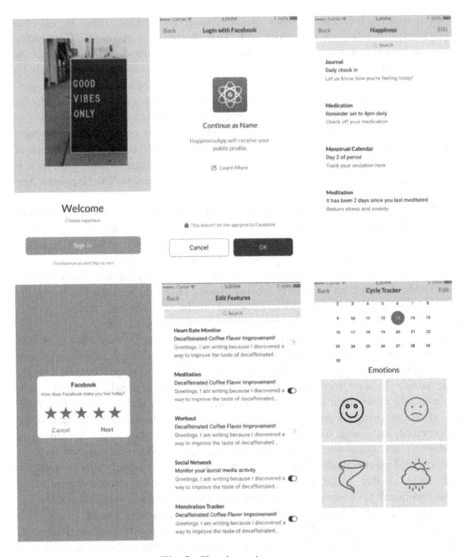

Fig. 8. Happiness App screens

While the app has a vast range of capabilities, it is easy for users to record, track and predict how to combat depression, anxiety and stress.

4 Discussion

As using social media has become an active part of our daily lives, it becomes a habit. Many feel short term gratification from appreciation by their peers, and others feel false fear when approval is not reached to a level one expects. Therefore, part of solving this

problem requires breaking the cycle of cue, routine and reward. There are different approaches for treating young women with DAS affected by social media. Cold turkey is one approach that some people use to break the habit cycle. However, it can be unrealistic as it is difficult for many to stay away from social media given that it is no longer just virtual places for people to communicate with friends, but important platforms for people to work. Unplugging in the short term is a good thing and healthy for everyone to do from time to time, but it may not be a feasible long term solution to many. Doing nothing is also a potential option. This however could reinforce the habit cycle. Moderation is a more realistic way as it normalizes and decreases the amount of interaction that one might attach to social media. This is similar to using CBT as it is about changing the way one thinks, feels and interprets a situation. The Happiness App aligns with the moderation approach.

The negative consequences of social media on young women poses the question of the level of governance and responsibility on all, micro, meso and macro levels. At the micro level, the individual needs to take responsibility for their actions. This means posting cautiously and commenting cautiously. It is also up to the individual to call others out on their social media activity and take accountability for their actions online. Individuals need to learn to identify what is causing DAS and what the best method is to stop the negative feelings towards social media. Friends and family are also influential in ensuring young women around them do not feel DAS.

At the meso level, celebrities need to take responsibility for their actions as their posts and photos could have a direct influence on the public, setting a standard of what is fashionable and trending. But influencer on Instagram, for instance, could be a bigger issue simply because they are not being treated as conventional celebrities. Their impact could cause more negative influence when they are seen as "normal" people. The photo shopping of photos among celebrities and influencers creates unrealistic expectations for young women as to what sort of body shape they should have. Businesses are using social media to promote their products as another form of advertising. Traditional advertising platforms are known to negatively affect a women's body image. Today, social media is become a significant platform that is reinforcing these traditional advertising ideas. Instead of building a business model on unhappiness, businesses should promote a healthy body image to young women.

At the macro level, websites and social media companies such as Facebook, Instagram, Google, Weibo and WeChat have a responsibility to ensure their environment is safe for users. Facebook now has a suicide algorithm to track down words associated with suicide. Instagram has a notification screen asking the person if they are okay when they search about suicide and depression. Groups such as SAMSHA operate their suicide prevention lifeline on Twitter, Facebook and YouTube to educate and help individuals. Many governments, like New Zealand's, do not have control over social media companies from overseas. That is why it is so important for these social media companies to be held responsible globally.

In addition, the micro, meso and macro levels are intermeshed and affect each other in a social media ecosystem. We no longer just live in the physical world. It is clear for the age group of 16–24 years old, the virtual world has become a large (if not a greater) part of their lives.

5 Conclusion

The current procedural and technological responses towards depression, anxiety and stress do not cater for young women. It is also apparent from the literature review that social media plays a contributing part towards young women's mental health. Following the design science research methodology, conceptual and system artefacts are identified. These artefacts highlight the habit loops that young women are in, and their desire for gratification through social media. Models were created showing the AS-IS and TO-BE of DAS. A proof of concept (Happiness App) was created that targets social media and young women affected by DAS. It demonstrates that the procedural and technological responses towards DAS are improved for young women as a more efficient app is available and targets their needs. This provides them with another channel to receive help and makes them aware of their own activities on social media. Finally, the micro, meso and macro levels of DAS among young women influenced by social media are discussed to illustrate that all levels of a society have their fair share of responsibility.

References

1. Mental Health Foundation: Quick facts and stats (2014). https://www.mentalhealth.org.nz/assets/Uploads/MHF-Quick-facts-and-stats-FINAL-2016.pdf
2. Mental Health – Helplines. https://www.mentalhealth.org.nz/get-help/in-crisis/helplines/
3. How stress affects your body and behavior. https://www.mayoclinic.org/healthy-lifestyle/stress-management/in-depth/stress-symptoms/art-20050987
4. Ministry of Health: Health Loss in New Zealand: A report from the New Zealand Burden of Diseases, Injuries and Risk Factors Study, 2006–2013. Ministry of Health, Wellington (2016)
5. McManus, S., Bebbington, P., Jenkins, R., Brugha, T.: Mental health and wellbeing in England: adult psychiatric morbidity survey 2014 (2016)
6. Rao, U., Hammen, C., Daley, S.: Continuity of depression during the transition to adulthood: a 5-year longitudinal study of young women. J. Am. Acad. Child Adolesc. Psychiatry **38**, 908–915 (1999)
7. Hintz, S., Frazier, P., Meredith, L.: Evaluating an online stress management intervention for college students. J. Couns. Psychol. **62**, 137–147 (2015)
8. Bewick, B., Gill, J., Mulhern, B., Barkham, M., Hill, A.: Using electronic surveying to assess psychological distress within the UK student population: a multi-site pilot investigation. E-J. Appl. Psychol. **4**, 1–5 (2008)
9. Scott, K., Oakley Browne, M., Mcgee, M., Elisabeth Wells, J.: Mental-physical comorbidity in te rau hinengaro: the New Zealand mental health survey. Aust. NZ J. Psychiatry **40**, 882–888 (2006)
10. Perrin, A., Anderson, M.: Share of U.S. adults using social media, including Facebook, is mostly unchanged since 2018. https://www.pewresearch.org/fact-tank/2019/04/10/share-of-u-s-adults-using-social-media-including-facebook-is-mostly-unchanged-since-2018/
11. Simoncic, T., Kuhlman, K., Vargas, I., Houchins, S., Lopez-Duran, N.: Facebook use and depressive symptomatology: investigating the role of neuroticism and extraversion in youth. Comput. Hum. Behav. **40**, 1–5 (2014)

12. O'Keefe, G., Clarke-Pearson, K.: The impact of social media on children, adolescents, and families. Pediatrics **127**, 800–804 (2011)
13. Fox, J., Moreland, J.: The dark side of social networking sites: an exploration of the relational and psychological stressors associated with Facebook use and affordances. Comput. Hum. Behav. **45**, 168–176 (2015)
14. Abel, J., Buff, C., Burr, S.: Social media and the fear of missing out: scale development and assessment. J. Bus. Econ. Res. (JBER) **14**, 33–44 (2016)
15. Fullerton, J.: Teenage girl kills herself 'after Instagram poll' in Malaysia. https://www.theguardian.com/world/2019/may/15/teenage-girl-kills-herself-after-instagram-poll-in-malaysia?CMP=Share_iOSApp_Other
16. Brenneisen, N.: The teens using instagram hashtags to glorify suicide. https://www.vice.com/en_uk/article/9bg7w8/glorifying-mental-illness-on-instagram-876
17. Depression Anxiety Stress Scales – DASS. http://www2.psy.unsw.edu.au/dass/
18. Jordan, P., Shedden-Mora, M.C., LoÈwe, B.: Psychometric analysis of the generalized anxiety disorder scale (GAD-7) in primary care using modern item response theory. PLoS One **12**, 1–14 (2017)
19. Thompson, E.: Hamilton rating scale for anxiety (HAM-A). Occup. Med. **65**, 601 (2015)
20. Wilhem, K., Reddy, J., Crawford, J., Robins, L., Campbell, L., Proudfoot, J.: The importance of screening for mild depression in adults with diabetes. Transl. Biomed. **08**, 1–8 (2017)
21. Antidepressant medication. https://www.sane.org/information-stories/facts-and-guides/antidepressant-medication/
22. Survey shows one in three adults with common mental disorders report using treatment services - NHS Digital. https://digital.nhs.uk/news-and-events/news-archive/2016-news-archive/survey-shows-one-in-three-adults-with-common-mental-disorders-report-using-treatment-services/
23. Beyond Blue. https://www.beyondblue.org.au/the-facts/depression/treatments-for-depression/medical-treatments-for-depression
24. Harris, R.: ACT Made Simple, 2nd edn. New Harbinger Publications, Oakland (2019)
25. Merry, S., Stasiak, K., Shepherd, M., Frampton, C., Fleming, T., Lucassen, M.: The effectiveness of SPARX, a computerised self-help intervention for adolescents seeking help for depression: randomised controlled non-inferiority trial. BMJ **344**, e2598 (2012)
26. Trost, Z., Parsons, T.: Beyond distraction: virtual reality graded exposure therapy as treatment for pain-related fear and disability in chronic pain. J. Appl. Biobehav. Res. **19**, 106–126 (2014)
27. Hevner, A.R., March, S.T., Park, J., Ram, S.: Design science in information systems research. MIS Q. **28**, 75–105 (2004)
28. Duhigg, C.: The Power of Habit: Why We Do What We Do in Life and Business. New York Random House, New York (2012)
29. Elhai, J.D., Hall, B.J., Erwin, M.C.: Emotion regulation's relationships with depression, anxiety and stress due to imagined smartphone and social media loss. Psychiatry Res. **261**, 28–34 (2018)
30. Brailovskaia, J., Teismann, T., Margraf, J.: Physical activity mediates the association between daily stress and Facebook Addiction Disorder (FAD) – a longitudinal approach among German students. Comput. Hum. Behav. **86**, 199–204 (2018)

Security, Privacy and Trust

Attainable Hacks on Keystore Files in Ethereum Wallets—A Systematic Analysis

Purathani Praitheeshan$^{(\boxtimes)}$, Yi Wei Xin, Lei Pan, and Robin Doss

School of IT, Deakin University, Geelong, VIC 3220, Australia
{ppraithe,ywxin,l.pan,robin.doss}@deakin.edu.au

Abstract. Ethereum is a popular Blockchain platform that allows users to manage their cryptocurrency transactions through the wallets. Ethereum wallet helps interact with the blockchain network easily, and it keeps Ethereum cryptocurrency (Ether) transaction data of its users. The use of Ethereum and wallets grows rapidly. Since they handle huge value of crypto assets, attackers are keen to hack and steal Ethers from Ethereum wallets. But there lacks comprehensive security analysis, especially on keystore files in Ethereum wallets. There were a few incidents occurred with huge loss of Ethers in Etheruem wallets within the last five years. In this paper, we conducted a systematic analysis on hacking methods from the existing literature and conducted experiments to find how the Ethereum wallet's keystore file is vulnerable to the adversaries. Since the keystore file is secured with a password, we have used the brute-force and the dictionary attack to crack the password of the keystore file in Ethereum wallets. Our results showed that the dictionary attack is more efficient to hack the keystore file than the brute-force attack. Further, the keystore file is less vulnerable, if it is used complex password credentials.

Keywords: Ethereum · Wallets · Keystore · Security · Attacks

1 Introduction

Ethereum [12] is one of the well-known blockchain platforms which allows cryptocurrency transactions and decentralized applications (Dapp) in many use cases. Wallet is a prominent element that allows the users to manage their accounts and crypto-assets securely by connecting to the blockchain network [14]. This paper surveys the existing Ethereum wallets [1], their functions, privacy and how can they vulnerable to specific attacks. Since the value of digital currency held in these wallets are increasing drastically, the attackers are keen to do malicious actions on wallets to gain substantial benefit. The recent attack exploited the parity multisig wallet [10] in Ethereum and the attackers stole around 150,000 Ethers [13,49] in 2017. Ether is the programmable token and

© Springer Nature Switzerland AG 2019
R. Doss et al. (Eds.): FNSS 2019, CCIS 1113, pp. 99–117, 2019.
https://doi.org/10.1007/978-3-030-34353-8_7

used as cryptocurrency in Ethereum [65]. People are performing their transactions using Ethers, and they use wallets as a storage medium [30] on Ethereum platform. Therefore, the level of security and privacy of Ethereum wallets should be ensured for the users who would trust the wallets functionalities.

It is crucial to investigate the security and privacy issues existed among the Ethereum wallets applications. There are a few literature surveys conducted on analysing security, performance, and privacy of Ethereum and other platforms [17,23,42,43,60]. However, no existing surveys focus on the Ethereum wallets and their keystore files. The user accounts in Ethereum wallets have their own keystore files that are authenticating with a secured password. We found the research gap in related to the password authentication in keystore files of Ethereum wallets and our research questions are generalized as following:

- What is the state of the art in the current attacks and countermeasures on keystore files in Ethereum wallets?
- How effective are the existing attacks breaching the passwords of Ethereum wallets keystore files?
- What kinds of new attacks will emerge in the future?

To address our research questions, we have conducted a systematic survey on the existing Ethereum wallets and their actual keystore files that are authenticating the investors/users of the distributed applications (Dapp) [16]. We mainly focus on the emerging security issues in Ethereum wallets, especially on the client-side. Ethereum wallet keeps a `keystore` file for each account and stores user's authentication data in JSON format [32]. Our analysis is primarily based on the `keystore` file that is the most important element for the hackers to steal the authentication information of Ethereum wallet users. We have done experiments using two password hacking mechanisms that are the `brute-force` and the `dictionary` attacks.

The remainder of the paper is organized as follows. Section 2 describes different types of Ethereum wallets and how `keystore` file is generated with secured authentication parameters. A literature review is presented in Sect. 3 that covers the possible hacking mechanisms and vulnerabilities in the platforms where the Ethereum wallets are running on. Section 4 presents our experiments using two different hacking methods and how the password of `keystore` file is cracked. The test results and evaluations on our methodologies are explained briefly in Sect. 5. Section 6 discusses a few other attacks that would be possible to target Ethereum wallets in future and concludes the paper.

2   Ethereum Wallets

Ethereum wallets are used by the owners of Ethereum accounts who save the private key and the public key of their accounts in a `keystore` file. These keys are authenticated when a user invokes or signs a transaction to manage their funds. The `hot wallets` store the key information online which can be accessible virtually from wherever the account holder has an internet connection. These

wallets are more risky to leak the key data since malicious actions can be easily processed in the cyber space to hack keys and steal money. The `cold wallets` keep the keys offline such as on a hardware device or in a paper wallet in the form of a QR code. The cold wallets are less vulnerable to attackers.

Ethereum wallets connect to the blockchain network fully or partially according to their configurations as either a full node or not. There are different types of Ethereum wallets which work on multiple platforms including desktop, mobile, web, and hardware devices.

Full Node Wallets/Desktop Wallet download the entire blockchain data and connect locally to operate their functions. Full node means one of the peers in the whole distributed Ethereum network. Since the blockchain data are very big (gigabytes in size), it is possible for a desktop wallet to run as a full node. It updates the state of transaction data in the blockchain by reaching an agreement with other nodes using the consensus mechanism of Ethereum.

Mobile Wallets do not require to acquire the whole blockchain data to connect with. They are light clients that they download only the miners who are necessary to send/receive transaction data whenever they invoke transactions. Since these wallets are installed in the mobile devices, it is easy to connect to the blockchain network any time or anywhere as long as the mobile has an internet access.

Web Wallets are online wallets that can be accessed from anywhere since wallet data are stored in a cloud environment. Since web wallets are light weight clients, they perform faster than other wallets. The online wallets are hot wallets, so that the accounts keys are stored in cloud. These wallets are most susceptible to malicious hacks and cipher attacks.

2.1 Offline Wallets

Are storing data and key information on their specially designed hardware storage or as the paper code, for instance, QR code. They generate the account keys offline and connect to the Ethereum network whenever the user needs to proceed their transactions. Since they are cold wallets, they are very resilient to the hackers or malicious users to steal the keys and cryto-funds. Although the offline wallets are cold wallets, they will expose to the online environment while being used [68].

2.2 Keystore File in Ethereum Wallets

A `keystore` file is a confidential and authentication factor that stores account credentials of an Ethereum wallet's user. Different Ethereum wallets store their keys using different cryptographic methods. They use either single public key or multiple private keys. An example `keystore file` created by a real Ethereum wallet (Mist [5]) is shown in the following code snippet.

```
{
"address":"827413828a671294f10769c2acf4327a53fb9191",
"crypto":{
    "cipher":"aes-128-ctr",
    "ciphertext":"0549d206b8712115e2c8fc7a9a5d43b43b72817a5457e2c
                7e2ba9d7cd927c41a",
    "cipherparams":{"iv":"a911db13a903c3101d357dd97cc0c030"},
    "kdf":"scrypt",
    "kdfparams":{
        "dklen":32,
        "n":262144,
        "p":1,
        "r":8,
        "salt":"ec267231ced9db706a47cf12ec55c4b1323b404488e1b15
                5c67342a78dfd3ff"
    },
    "mac":"9aaaf70488208a3c007a167310ed9900137f81bc07f1cad72ade6
          43628cb9e3d"
    },
    "id":"f6e0435d-b575-4c42-aa13-a38cfb857869",
    "version":3
}
```

The keystore file has the important key-value pairs that store the security information of the private key, public key, id, addresses, and cipher parameters. We explain them in details—*address*: Address of the Ethereum wallet account that is used for transfer and receive Ethers. *cipher*: The name of the Advanced Encryption Standard (AES) algorithm used. *cipherparams*: The parameters required by the *cipher* algorithm. *ciphertext*: The private key of Ethereum, encrypted by using *cipher* algorithm. *kdf*: Key derivation function used by Ethereum wallet, which enables the end user to encrypt the keystore file with password. *mac*: Message authentication code used to verify the correctness of user password.

A keystore file name consists of a timezone indicator (e.g.,UTC), date, time, and the Ethreum address information as UTC--2019-04-17T02-24-31.54696700 0Z--827413828a671294f10769c2acf4327a53fb9191. The file would be saved in chaindata/keystore path inside the private or public blockchain data folder, if the wallet is installed on a computer.

Since these keystore files are stored in plain text, it is easy to obtain both the file and the content of the file. If the system has not enabled with any encryption protection such as operating system level encryption or full disk encryption, the attackers can try malicious actions to retrieve the keystore file from the exact file path. If this keystore file is leaked to hackers, they can use different password hacking mechanisms to reveal the actual password of the Ethereum wallet.

Once the address of an Ethereum Wallet account is leaked, it can be used to trace more important information, including the transaction information, wallet status, and account balance. There are online scanners such as Etherscan [4] and EtherChain [2], freely available to explore and trace an Ethereum address with its transaction history information. Transaction data such as transaction time, amount of Ethers, and the address to whom the money transferred or received can be searched by the scanners.

If a keystore file is retrieved by an attacker, the hashcat application [7] can be used to crack the password of the keystore file [33,34]. Both the dictionary attack and the brute-force attack are used to hack the password of Ethereum wallet using private key stored in the keystore file. With the password information, the hacker can gain the full control of the Ethereum wallet.

3   Literature Review on Hacking Mechanisms

There are many researchers contributed their work in respect to password hacking mechanisms. Such methods can be used to crack password of Ethereum wallet with the data retrieved from the keystore file.

3.1   Computers/PC Related Attacks

Gelernter et al. [27] introduced the Password Reset Man in the Middle attack (PRMitM). This method is derived from the traditional Man-in-the-Middle (MitM) attack, and it can affect popular websites and online services including email services. This attack works by collecting users' personal details while they register by tricking the victims to provide their personal information. This data can be used to reset password for another online service either via email, SMS, or phone call. There were two proposed countermeasures that force users to understand when someone asked to reset their passwords.

Naiakshina et al. [48] conducted a qualitative research on how developers introduced problems with password storage on applications, websites, etc. They have identified different types of reason, thoughts, and actions which will weaken password in applications. The experiment was conducted in a laboratory environment with the Java programming language. This research was conducted through studying the computer science students instead of the real developers, so that the results were limited.

Ge et al. [26] conducted research on loader attacks. It was discovered that dynamic loading is one of the core features on modern operating systems. A new attack vector was identified as Copy Relocation Violation (CORev) where the adversaries can utilize a vulnerability of memory corruption to alter the read-only constant variables in order to bypass defenses. Ge et al. [26] suggested three mitigation methods—analyzing libraries and binaries to detect the attack vector, recompilation, and make loader and linker to be aware of permissions at the source level.

Silver et al. [53] focused on attacks and mitigations on password managers. They investigated the security issues of password managers and policies on automatic filling function in web passwords. It was discovered that a remote network adversary can extract multiple passwords from the password manager without any user interaction. Their experiments were done for improving the security of password managers. They presented a solution called `SecureFilling` that enhanced the security of auto-filling function of the password managers.

Kogan et al. [39] investigated the second-factor authentication which utilized secure hash chains. They used time-based one-time password (TOTP) that can store the secrets on both the server and the client. A new one time password scheme named `T/Key` was proposed that is time-based and offline. Independent hash functions were implemented in `T/Key` to force the server not to store any secrets of secondary authentication.

Song et al. [54] have used the behavioral information and hand geometry to authenticate the users on multi-touch enabled devices. Their research indicates that the traditional password and gesture pass-code are vulnerable against several attacks such as zero-effect attack, smudge attack, shoulder attack, and statistical attack. They developed a solution that uses the multi-touch authentication method. It can protect the victims from statistical attack, shoulder attack, and smudge attack effectively.

Su et al. [56] explored the crosstalk leakage attack on USB hubs. This attack can be monitored via the USB power cords. The USB charge-only cables, internal USB hubs and external off-the-shelf USB hubs would be affected by the crosstalk attack.

Xiao et al. [67] researched on utilizing a side channel to trace and detect SSL/TLS vulnerabilities in the secured enclaves. An enclave is Intel Software Guard Extension (SGX) which provides a shielded execution environment for the software applications. They proposed an analysis framework called `STACCO` that dynamically analyzes the SSL/TLS implementation and identifies vulnerabilities such as man-in-the-kernel attack, oracle attacks, and so on.

Tian et al. [59] focused on the provenance issue in data protection of USB storage devices. `ProvUSB` can be utilized to find grained provenance collections and track data storage problems on smart USB devices.

Wressnegger et al. [66] conducted research on vulnerabilities introduced when migrating systems to 64-bit platforms. They discussed that code works on 32-bit platforms can introduce vulnerabilities to 64-bit platforms. The predicted vulnerabilities are due to the differentiation on data models, integer types and widths on both platforms.

3.2   Mobile Related Attacks

Luo et al. [44] investigated the vulnerabilities existed in the User Interface (UI) of mobile browsers. They developed `Hindsight` which is the first framework of browser-agnostic testing and dynamic-analysis. It is quantifying and gauging the vulnerabilities of mobile web browsers specifically for UI attacks.

Li et al. [41] investigated mobile WebView that focuses on cross-app remote infections. The convenient cross-app URL invocation can be utilized to execute unauthorized app components. They found that cross-WebView navigation design allows the remote adversary to distribute malicious web content among different apps such as Google Drive, Twitter, Facebook, etc. Some mitigation techniques for this attack was discussed and proposed a security mechanism called `NaviGuard` to use within app.

Redini et al. [50] studied bootloader security in mobile devices and discovered that unlock ability can make bootloader vulnerable. `BOOTSTOMP` is their proposed solution that is a multi tag taint analysis framework to automatically identify the security vulnerabilities in the bootloader.

Genkin et al. [28] focused on the extraction of the ECDSA key from mobile devices by utilizing physical side channels. They demonstrated a full extraction process of the ECDSA secret signing keys from CoreBitcoin application and OpenSSL runs on iOS or Android mobile devices.

Hojjati et al. [31] conducted research on the side channel attack which revealed the factory floor secrets via mobile phone. They investigated possible attacks targeting factories and other trading secrets. They utilized the embedded sensor devices including GPS, camera, microphone and accelerometer to identify side channel attacks.

Das et al. [25] investigated the attacks and defenses of tracking the mobile web users via motion sensors. They analyzed the accelerometer and gyroscope motion sensors for fingerprinting and tracking the user. The research discovered that motion sensor tracking can have very high accuracy. Their proposed mitigation methods are related to the primary obfuscation techniques when explicit user permission/awareness is absent.

3.3   Networking Related Attacks

Chen et al. [21] systematically studied on MitM attack using name collision. They explained name collision problem that makes MitM attack on web browsing become easier. The paper mainly described the technical details of the related attacks, name collision and MitM attack.

Chen et al. [22] also conducted detailed research on client-side name collision vulnerability. They have discovered that the name collision problem can lead to MitM attacks against end-user devices on the internet. The research confirmed that attackers can register vulnerable domains with leaks from WPAD queries, and then the global web traffic from internet users can be automatically redirected to the attacker's MitM proxy.

Vissers et al. [63] had studied on domain hijacking via name-servers with a large-scale analysis. They described the email hijacking is another type of technique to be used to launch a domain hijacking. The research group discussed security practices of name-servers such as Domain Name System Security Extensions (DNSSEC), which can be used to protect DNS from integrity issues by utilizing digital signatures.

Vanhoef and Piessens [61] conducted research on the newly discovered WPA2 key re-installation attack which forces to reuse nonce on Wi-Fi. The researchers proposed the mitigation techniques and indicated that the vendors would be notified about this vulnerability and attack.

Jero et al. [36] focused on the attacks and defenses within software-defined network (SDN) from the identifier binding perspective. They discovered that the SDN is flawed and proved with their developed proof-of-concept attack by using SDNs. The experiments showed that their solution mitigates the identifier binding attacks at the cost of a little overhead.

Varadarajan et al. [62] worked on the placement vulnerability existing in public clouds. They primarily discussed the placement of virtual machines (VM) in the public clouds. Their research discovered that if the adversary VM was placed right, it can be used to launch a side-channel attack to the nearby victims. The research discussed about the detection of co-residence. Multi-tenancy in public clouds can have the co-residency attack issue; the deficient of performance isolation in hardware enables the detection of the co-location; and it is easy and cheap to achieve the co-location.

Zhang et al. [69] conducted research on side-channel attacks in PaaS clouds with cross-tenants. It presented a new attack framework which can be utilized to cache-based side-channel attacks on the Platform-as-a-Service (PaaS) clouds. They have utilized FLUSH-RELOAD framework to achieve the attack.

3.4 Web Related Attacks

Tajalizadehkhoob et al. [57] studied the website vulnerabilities and security issues in the view points of the web hosting provider or the website administrator. They stated that the web security in the shared hosting environment is joint responsibility between webmasters and providers.

Jin et al. [37] conducted research on the code injection problem that exists on HTML5 based mobile apps. They have studied about the portability advantage of HTML5 based mobile apps that is attracting more attention from developers. They have found a new form of code injection that inherits the basic cause of Cross-Site Scripting attack (XSS). To analyze the prevalence of this code injection vulnerability, they have developed a vulnerability detection tool that investigates PhoneGap apps.

Cao et al. [20] focused on timing attacks against web browser. Timing attacks pose threats to modern web browsers that lead to privacy and security threats. They discovered that the existing countermeasures are not sufficient. The researchers proposed a novel approach named deterministic browser which can mitigate timing attacks in modern browsers provably. Their prototype of deterministic browser named DETERFOX can protect victim from a few timing attacks and also compatible with the real-world websites.

Zuo et al. [70] conducted study on an automatic discovery of vulnerable authorizations existing within online service. The research discovered that many apps implemented login and authorization features for use of app. But they have different levels of security implementation which can propose different types of

threats to the end users and web servers. They have developed an automatic online service access control vulnerability detection system namely Authscope.

Shan et al. [52] conducted research on the tail attacks on web applications. The tail attack is the extension of Distributed Denial-of-Service (DDoS) attacks to the application layer. The adversary exploits a newly identified vulnerability in system of n-tier web applications.

Lauinger et al. [40] have done a comprehensive analysis on the outdated JavaScript libraries used on the web, specially in the client-side JavaScript library. They discovered that since JavaScript libraries such as Bootstrap, Angular and jQuery are frequently used on many websites, the attack surface is increasing.

Sanchez-Rola et al. [51] conducted research on the security of browser extension policies. The research discovered that due to the tight relation with browsers, browsers extensions have been the target of many attacks. These attacks are related to gather or steal information, execute malicious tasks at background, password theft, and browsing history retrieval etc. It also provided mitigation techniques against the defined attacks including a side-channel attack and a set of URI leakage security threats.

Han et al. [29] focused on live monitoring of controlled sandboxed phishing kits. They presented a new approach to sandbox real-time phishing kits which can protect the victims. They have designed a honeypot system by incorporating their module and the collected data. The researchers measured the effective lifetime of phishing kits with respect to the separated data of victims, adversaries, and other third-party victories.

Invernizzi et al. [35] conducted research on the detection of cloaked websites. A cloaking attack happens when an adversary creates a webpage with two versions of targeting search engine and targeting human. They have used their findings to develop an anti-cloaking system that detects split-view.

Muthukumaran et al. [47] conducted research on mitigation of data disclosure vulnerabilities in web applications. They discovered that bugs and logic of web applications related to authentication can expose the user data. The research discussed common defense techniques including access check, input validation, anomaly detection, policy, etc. With their deep analysis of existing vulnerabilities, they proposed a proxy named *FlowWatcher* that is used to mitigate the data disclosure vulnerabilities in web applications.

Meng et al. [45] studied on the pollution attack of targeted advertising. It presented a new fraud technique which enables the publishers to increase their advertisement income. They have discovered that the pollution attack utilizes many different techniques, for example, differentiating search engine and real users to avoid being blacklisted by search engines.

Monshizadeh et al. [46] conducted research on the detection of privilege escalation attacks in web applications. The research discovered that many websites have privilege escalation vulnerability. This problem can cause further damage or disclosing the important data, due to the complex structure of websites, including utilization of server-side script programs and database server.

Karapanos et al. [38] focused on the mitigation of TLS man-in-the-middle (MITM) attacks in web applications. It has explained the technical details with diagrams visualizing the concept of attack. It proposed a solution to use the dubbed Server Invariance with Strong Client Authentication (SISCA) in order to defend against user impersonation used in the TLS MITM attacks.

Soska et al. [55] conducted research on automating the detection of vulnerable websites before they become malicious. The approach is to classify the websites by utilizing data mining and machine learning techniques.

Canali et al. [19] studied on exploitation behaviors on the web. In order to analyze the exploitation behavior, they have presented the design, implementation and deployment of network with 500 fully functional honeypot websites with different services for attract attackers. It discovered that many eastern European countries were involved in spam comment, and many phishing and scam campaigns were operated by criminals located in African countries.

4   Methodology and Experiments

We have studied possible attack mechanisms from the existing literature surveys that would be directly or indirectly applicable for hacking Ethereum wallets. Our study also analyzed possible attacks that can be targeted to Ethereum wallets. This section describes our experiments done on cracking different private keys that we extracted from the `keystore` files in Ethereum wallets. To the best of our knowledge, this approach is the first analysis on security of Ethereum wallets with significant experiment results. We discovered that the prominent element is the `keystore` file that would be targeted in any attacks on Ethereum wallets. If an adversary obtained the keystore file and its password, he/she will have the full control of the particular Ethereum address/wallet.

Since an Ethereum wallet requires the minimum of 8 characters for the wallet password, we tested only the Ethereum keystore file with 8 characters password. For our experiments, the adversary machine had an Intel Core CPU with with 32GB RAM, and Windows 10 with the installation of the Hashcat application with the rate of 10-11 Hashes/s. The victim machine had an Intel Core CPU with 4GB RAM, and MacOS operating system. Hashcat [7] is a free and open-source multi-OS password cracker with in-kernel rule engine. It supports over 200 hash types and password/hash cracking by using Graphic Processing Unit (GPU) and Central Procession Unit (CPU). We used the Windows version of hashcat binary for our cracking experiment. The Python script `ethereum2john.py` [11] was used to convert the keystore file to a compatible format of hashcat. Hashcat has a few configuration parameters to set up initially. We used the specific values to enable Hashcat work properly with the Ethereum keystore files as shown in following Table 1.

4.1   Brute-Force Attack

The first hacking method we have chosen in this study is the brute-force attack. It tries various passwords repeatedly until it gets in successfully to the

Table 1. Hashcat configuration parameters and values

Parameters	Description
-m 15700	Switch for hash mode, 15700 SCRYPT hash type for wallet
–status	Switch to enable automatic update of the cracking progress status
–status-timer=5	Set update interval for updating progress
-D 1	Use CPU to calculate hash for cracking
-w 3	Workload profile switch, 3 means high priority
dictionary text file	Use specified dictionary to crack the hash
–potfile-disable	Disable potfile which used to store cracked passwords
-a 3	Attack mode brute-force
-1 ?l?u?d	User defined charset, lower case, upper case and digits
?1?1?1?1?1?1?1?1	8 characters
–increment	Switch to enable mask increment mode
–increment-min 8	Minimum 8 characters
–increment-max 8	Maximum 8 characters

wallet/system. As initial setup, we installed the tool `geth` (Ethereum protocol [6]) and Ethereum wallet (MyEtherWallet [9]) on MacOS. Then we created an account using Ethereum Wallet and it generated a keystore file in the `chaindata/keystore/` directory. The Hashcat version 4.2.1 [8] was used in our experiments for cracking the keystore file password. It was installed by executing the installation file using the command prompt. Next, we configured the Python environment to run `ethereum2john.py` script. This script file and the keystore file were copied into the python executable directory. The following command would execute the ethereum2john.py to convert the keystore file to a compatible hashcat format: `phython ethereum2john.py keystore-filename`.

For example, our keystore file name is UTC--2019-04-17T02-24-31.5469 6700 0Z--358d23c9b76d113260ab7288ea8ffbd8d8a3f9f3, and the executed command is as follows inside the python executable directory.

    python ethereum2john.py UTC--2019-04-17T02-24-31.546967000Z--35
8d23c9b76d113260ab7288ea8ffbd8d8a3f9f3

The following output shows the result we received from above command. It is the hash format of the keystore file that to be compatible to run in the hashcat software.

```
WARNING: Upon successful password recovery, this hash format may
expose your PRIVATE KEY. Do not share extracted hashes with any
untrusted parties!
UTC--2019-04-17T02-24-31.546967000Z--358d23c9b76d113260ab7288ea8ffb
d8d8a3f9f3:$ethereum$s*262144*8*1*0aead560af9504a38f7bc500cf26b0588
cc30a094bdd7320597ff9cb6154c4fd*5029816b5bae91586929e7132d2c679bc9e
6187b99154973f05ca56166a1b80b*c3947083ae6b4e628753d9b6d20fefa3a62eb
1c3b14cd9dca3033afe032bd1ef
```

Cracking. The hashcat executable file was run to extract the keystore file from the output we received using the `ethereu2john.py` script. The following command is used for cracking the keystore file.

```
hashcat64.exe -m15700 $ethereum$s*262144*8*1*0aead560af9
504a38f7bc500cf26b0588cc30a094bdd7320597ff9cb6154c4fd*5029816b5bae9
1586929e7132d2c679bc9e6187b99154973f05ca56166a1b80b*c3947083ae6b4e6
28753d9b6d20fefa3a62eb1c3b14cd9dca3033afe032bd1ef --status
--status timer=5 -D1 -w1 -a3
```

The above command extracted the keystore file using hashcat cracking algorithm. It used the brute-force attack as specified in the command as -a3 mode.

4.2   Dictionary Attack

The second hacking mechanism we experimented to crack an Ethereum wallet keystore file is the dictionary attack. We trimmed the `rockyou` dictionary [3] so that the passwords with only 8 characters are kept.

Cracking. We used the hashcat tool [7] with the dictionary file to crack the passwords using hashcat dictionary mode. The following command is for cracking the keystore file using a dictionary file and hashcat executable file.

```
hashcat64.exe -m15700 $ethereum$s*262144*8*1*0aead560af9
504a38f7bc500cf26b0588cc30a094bdd7320597ff9cb6154c4fd*5029816b5bae9
1586929e7132d2c679bc9e6187b99154973f05ca56166a1b80b*c3947083ae6b4e6
28753d9b6d20fefa3a62eb1c3b14cd9dca3033afe032bd1ef --status
--status timer=5 -D1 -w3 ry8.txt --potfile-disable
```

4.3   Test Cases

We have conducted experiments with the keystore file generated in Ethereum wallet to investigate the possibilities of hacking methods to crack the wallet credentials. In our tests, we chose 8 different passwords with the length of 8 characters. Since the minimum password length of Ethereum wallet is eight, we assumed that most of the users use 8 characters-long passwords for their wallet accounts. The selected passwords that we used in our test cases are 00000000, 12345678, a1234567, abcdefgh, kangaroo, 10294538, anchdksl and aSjk39fD. The reasons that we have chosen these passwords are explained in Table 2. The keystore files with these passwords are available to download from Github [3].

After passwords were chosen, we tested both brute-force attack and dictionary attack using hashcat to crack the selected keystore files. For each password and both attacks, we ran cracking 40 times in order to get reliable and accurate result. We have set the maximum cracking time to one hour for both attacks scenarios to check weather the keystore file is successfully cracked or failed. Further, we used the masked cracking feature in hashcat that sets up more specific

Table 2. Selected passwords and reasons for choosing them [18,64]

Password	Reason to choose
00000000	Users usually use same characters with minimum length
12345678	Users use continues digits to memorize easily
a1234567	Users use very simple combination of alphabet and digits with minimum length
abcdefgh	Users use continues alphabetical letters to memorize easily
kangaroo	Users use their favourite word with the minimum length and easy to memorize
10294538	Advanced users use password generator to generate random password in digits
anchdksl	Advanced users use password generator to generate random password in letters
aSjk39fD	Advanced users use password generator to generate random passwords include upper, lower case letters and digits

rules to brute-force attack. The masked cracking was increased the success rate and reduced the cracking time [58]. The results from all the test cases with each password are uploaded in Github repository [3] for future research.

5   Problem Analysis and Discussion

In our experiments, we assumed that an adversary was able to get the keystore file from the victims's machine that was generated by an Ethereum offline/desktop wallet as it is stored in the default directory `chaindata/keystore/`.

The challenges exist in Ethereum wallets are much related to the problems and attacks we discussed in Sect. 3, literature review. For an example, Wi-Fi key re-installation attack combined with phishing attack can lead to leak the Ethereum wallet's password and the keystore file as well. An adversary can pretend to be the public Wi-Fi, and send request to download file to the victim with fake portal page. It will request the victim to download a program to connect to internet. This method allows attacker to trick victims to download Trojan horse that would steal the Ethereum wallet keystore file and password.

In our study, we have analysed the possibilities of two hacking mechanisms that are brute-force and dictionary attack. The result from brute-force attack showed that Hashcat is failed to crack all of keystore files within 1 hour. It is estimated at least to spend more than ten years to crack the password by trying every combination of lower case, upper case and digits. It makes brute-force so hard if the user used complex password using letters and digits. The mask feature in hashcat allowed us to specify the format of password for brute-force cracking. For example, we can define the values for length, character or numeric values for

a password to be cracked. Adding mask in brute-force attack was able to crack reduced the estimated time to 116 days and passwords cracked within 13 min. The success rate was increased and the passwords of 00000000 and 12345678 were cracked successfully. Other passwords were failed to crack even with the mask brute-force condition.

In the dictionary attack, weak passwords with single characters, continues digits or letters, combination of simple digits and letters and common words like kangaroo were easy to crack since they are listed in the rockyou dictionary file. The cracking time for password 12345678 was very less compare to the time taken for password a1234567. Interestingly password 00000000 was not the fastest to be cracked even it is a simple one. The cracking efficiency is depends on where the password is located in the dictionary file. If we look up the dictionary file, it was observed that the password resides in top lines were cracked with less time than the password that are in bottom lines in the files. Further we discovered that random digits, random characters and random mixture of upper case, lower case letters and digits can help to increase the level of complexity. Hashcat was unable to crack the keystore file with password 10294538, anchdlksl and aSjk39fD within an hour. The common and weak passwords have higher chance to appear in the dictionary file and more vulnerable to dictionary attacks. Random passwords have higher complexity level and they are not in the dictionary file. Thus, the chance of cracking passwords that includes random letters and digits is less compare to the common passwords that uses continuous digits/letters and easy words.

Overall, our experiment results showed that the dictionary attack on Ethereum keystore files uses less time to crack the password. But it also failed to crack complex password within the expected time. The brute-force attack is not a good hacking mechanism to crack Ethereum keystore file since it was failed to crack all of the passwords which we selected. But it was able to crack some simple/weak passwords successfully when the masked brute-force was applied.

With our study, we investigated that the security of Ethereum wallet keystore file is good enough to keep apart from adversaries. The major hacking factor of keystore file in Ethereum wallet is the complexity. From our experiments, we concluded that the complex passwords are more secured than common or simple passwords. We also discovered that the Ethereum wallet is using SCRYPT [15, 24] that enables the password cracking very difficult due to its memory hard hash algorithm.

6   Conclusion

This paper reviewed the state of the art of attacks which can be utilized to attack Ethereum wallets and analyzed the experiment we have conducted with brute-force and dictionary attack in the Ethereum wallet keystore file. We investigated that the main goal of adversaries is to steal the keystore file which is generated by Ethereum wallet to drain off Ethers from the victim's account. This keystore file contains very important and sensitive information about the Ethereum wallet.

There is no encryption applied to the keystore file and it is stored in a plain text file with content in JSON format. Any malicious user who has access to the Ethereum wallet file directory, can steal this file and use hashcat to crack its password.

In the future, there are possibilities that adversaries can utilize different and combinations of attack methods to steal Ethers from victim. They can combine WiFi key re-installation with phishing attack and click bait attack to obtain passwords from the victim directly. The keystore file also can be stolen by utilizing trojan horse and fake websites. The attackers can utilize drive by downloads to implant trojan horse to victim's device and get the keystore file. The adversaries are able to utilize search engines to do profiling of some Ethereum wallets and then prepare for spear phishing attack combined with click baits to find keystore file and password.

To improve the security of Ethereum wallet, it is necessary for account holders to use more complex passwords, such as those password must include combination of upper case, lower case characters, digits, special characters and the minimum length of password must be 8. Further, it is advisable to encrypt the keystore file instead of saving it in a plain text form on the disk. These awareness methods would help to make the Ethereum wallet well secured.

References

1. Blockchain platform: Ethereum. https://www.ethereum.org/
2. Etherchain - The Ethereum Blockchain Explorer. https://www.etherchain.org/
3. Ethereum Wallet Attacks and Countermeasure Assnalysis. https://github.com/coddec/ethereum-attack-countermeasure/tree/master/
4. Etherscan - The Ethereum Blockchain Explorer. https://etherscan.io/
5. Geth - The Go Implementation of Ethereum Protocol. https://github.com/ethereum/mist/
6. Geth - The Go Implementation of Ethereum Protocol. https://geth.ethereum.org/
7. Hashcat - An advanced password recovery tool. https://hashcat.net/hashcat/
8. Hashcat 4.2.1.7 - Download Software files. https://hashcat.net/files/hashcat-4.2.1.7z
9. MyEtherWallet - The Ethereum Original Wallet. https://www.myetherwallet.com/
10. Parity Wallet Library. https://github.com/paritytech/parity/blob/4d08e7b0aec46443bf26547b17d10cb302672835/js/src/contracts/snippets/enhanced-wallet.sol
11. The python script file to convert keystore file to hashcat compatible format. https://github.com/magnumripper/JohnTheRipper/blob/bleeding-jumbo/run/ethereum2john.py
12. Ethereum Foundation. Ethereum's white paper (2014). https://github.com/ethereum/wiki/wiki/White-Paper
13. An In-Depth Look at the Parity Multisig Bug (2016). http://hackingdistributed.com/2017/07/22/deep-dive-parity-bug/
14. Abe, J.: Bitcoin, wallet management and network security management with storage components: a model (2018)

15. Alwen, J., Chen, B., Pietrzak, K., Reyzin, L., Tessaro, S.: Scrypt is maximally memory-hard. In: Coron, J.-S., Nielsen, J.B. (eds.) EUROCRYPT 2017. LNCS, vol. 10212, pp. 33–62. Springer, Cham (2017). https://doi.org/10.1007/978-3-319-56617-7_2

16. Antonopoulos, A.M., Wood, G.: Mastering Ethereum: Building Smart Contracts and DApps. O'Reilly Media, Sebastopol (2018)

17. Atzei, N., Bartoletti, M., Cimoli, T.: A survey of attacks on ethereum smart contracts (SoK). In: Maffei, M., Ryan, M. (eds.) POST 2017. LNCS, vol. 10204, pp. 164–186. Springer, Heidelberg (2017). https://doi.org/10.1007/978-3-662-54455-6_8

18. Campbell, J., Ma, W., Kleeman, D.: Impact of restrictive composition policy on user password choices. Behav. Inf. Technol. **30**(3), 379–388 (2011)

19. Canali, D., Balzarotti, D.: Behind the scenes of online attacks: an analysis of exploitation behaviors on the web. In: 20th Annual Network & Distributed System Security Symposium (NDSS 2013) (2013)

20. Cao, Y., Chen, Z., Li, S., Wu, S.: Deterministic browser. In: Proceedings of the 2017 ACM SIGSAC Conference on Computer and Communications Security, pp. 163–178. ACM (2017)

21. Chen, Q.A., Osterweil, E., Thomas, M., Mao, Z.M.: MitM attack by name collision: cause analysis and vulnerability assessment in the new gTLD era. In: 2016 IEEE Symposium on Security and Privacy (SP), pp. 675–690. IEEE (2016)

22. Chen, Q.A., Thomas, M., Osterweil, E., Cao, Y., You, J., Mao, Z.M.: Client-side name collision vulnerability in the new gTLD era: a systematic study. In: Proceedings of the 2017 ACM SIGSAC Conference on Computer and Communications Security, pp. 941–956. ACM (2017)

23. Chen, T., et al.: Understanding ethereum via graph analysis. In: Proceedings of INFOCOM (2018)

24. Dannen, C.: Introducing Ethereum and Solidity. Springer, Heidelberg (2017). https://doi.org/10.1007/978-1-4842-2535-6

25. Das, A., Borisov, N., Caesar, M.: Tracking mobile web users through motion sensors: attacks and defenses. In: NDSS (2016)

26. Ge, X., Payer, M., Jaeger, T.: An evil copy: how the loader betrays you. In: NDSS (2017)

27. Gelernter, N., Kalma, S., Magnezi, B., Porcilan, H.: The password reset MitM attack. In: 2017 IEEE Symposium on Security and Privacy (SP), pp. 251–267. IEEE (2017)

28. Genkin, D., Pachmanov, L., Pipman, I., Tromer, E., Yarom, Y.: ECDSA key extraction from mobile devices via nonintrusive physical side channels. In: Proceedings of the 2016 ACM SIGSAC Conference on Computer and Communications Security, pp. 1626–1638. ACM (2016)

29. Han, X., Kheir, N., Balzarotti, D.: Phisheye: live monitoring of sandboxed phishing kits. In: Proceedings of the 2016 ACM SIGSAC Conference on Computer and Communications Security, pp. 1402–1413. ACM (2016)

30. He, S., et al.: A social-network-based cryptocurrency wallet-management scheme. IEEE Access **6**, 7654–7663 (2018)

31. Hojjati, A., et al.: Leave your phone at the door: side channels that reveal factory floor secrets. In: Proceedings of the 2016 ACM SIGSAC Conference on Computer and Communications Security, pp. 883–894. ACM (2016)

32. Homoliak, I., Breitenbacher, D., Binder, A., Szalachowski, P.: An air-gapped 2-factor authentication for smart-contract wallets (2018). https://doi.org/10.13140/RG.2.2.11358.69445

33. Houshmand, S., Aggarwal, S., Flood, R.: Next Gen PCFG password cracking. IEEE Trans. Inf. Forensics Secur. **10**(8), 1776–1791 (2015)
34. Hranický, R., Zobal, L., Ryšavý, O., Kolář, D.: Distributed password cracking with BOINC and hashcat. Digit. Investig. **30**, 161–172 (2019)
35. Invernizzi, L., Thomas, K., Kapravelos, A., Comanescu, O., Picod, J.M., Bursztein, E.: Cloak of visibility: detecting when machines browse a different web. In: 2016 IEEE Symposium on Security and Privacy (SP), pp. 743–758. IEEE (2016)
36. Jero, S., Koch, W., Skowyra, R., Okhravi, H., Nita-Rotaru, C., Bigelow, D.: Identifier binding attacks and defenses in software-defined networks. In: 26th {USENIX} Security Symposium ({USENIX} Security 2017), pp. 415–432 (2017)
37. Jin, X., Hu, X., Ying, K., Du, W., Yin, H., Peri, G.N.: Code injection attacks on HTML5-based mobile apps: characterization, detection and mitigation. In: Proceedings of the 2014 ACM SIGSAC Conference on Computer and Communications Security, pp. 66–77. ACM (2014)
38. Karapanos, N., Capkun, S.: On the effective prevention of {TLS} man-in-the-middle attacks in web applications. In: 23rd {USENIX} Security Symposium ({USENIX} Security 2014), pp. 671–686 (2014)
39. Kogan, D., Manohar, N., Boneh, D.: T/key: second-factor authentication from secure hash chains. In: Proceedings of the 2017 ACM SIGSAC Conference on Computer and Communications Security, pp. 983–999. ACM (2017)
40. Lauinger, T., Chaabane, A., Arshad, S., Robertson, W., Wilson, C., Kirda, E.: Thou shalt not depend on me: analysing the use of outdated javascript libraries on the web. arXiv preprint arXiv:1811.00918 (2018)
41. Li, T., et al.: Unleashing the walking dead: understanding cross-app remote infections on mobile webviews. In: Proceedings of the 2017 ACM SIGSAC Conference on Computer and Communications Security, pp. 829–844. ACM (2017)
42. Li, X., Jiang, P., Chen, T., Luo, X., Wen, Q.: A survey on the security of blockchain systems. Futur. Gener. Comput. Syst. (2017)
43. Lin, I.C., Liao, T.C.: A survey of blockchain security issues and challenges. IJ Netw. Secur. **19**(5), 653–659 (2017)
44. Luo, M., Starov, O., Honarmand, N., Nikiforakis, N.: Hindsight: understanding the evolution of UI vulnerabilities in mobile browsers. In: Proceedings of the 2017 ACM SIGSAC Conference on Computer and Communications Security, pp. 149–162. ACM (2017)
45. Meng, W., Xing, X., Sheth, A., Weinsberg, U., Lee, W.: Your online interests: Pwned! a pollution attack against targeted advertising. In: Proceedings of the 2014 ACM SIGSAC Conference on Computer and Communications Security, pp. 129–140. ACM (2014)
46. Monshizadeh, M., Naldurg, P., Venkatakrishnan, V.: MACE: detecting privilege escalation vulnerabilities in web applications. In: Proceedings of the 2014 ACM SIGSAC Conference on Computer and Communications Security, pp. 690–701. ACM (2014)
47. Muthukumaran, D., O'Keeffe, D., Priebe, C., Eyers, D., Shand, B., Pietzuch, P.: Flowwatcher: defending against data disclosure vulnerabilities in web applications. In: Proceedings of the 22nd ACM SIGSAC Conference on Computer and Communications Security, pp. 603–615. ACM (2015)
48. Naiakshina, A., Danilova, A., Tiefenau, C., Herzog, M., Dechand, S., Smith, M.: Why do developers get password storage wrong?: a qualitative usability study. In: Proceedings of the 2017 ACM SIGSAC Conference on Computer and Communications Security, pp. 311–328. ACM (2017)

49. Palladino, S.: The parity wallet hack explained, July 2017. https://blog.zeppelin.solutions/on-the-parity-wallet-multisig-hack-405a8c12e8f7
50. Redini, N., et al.: Bootstomp: on the security of bootloaders in mobile devices. In: 26th {USENIX} Security Symposium ({USENIX} Security 2017), pp. 781–798 (2017)
51. Sanchez-Rola, I., Santos, I., Balzarotti, D.: Extension breakdown: security analysis of browsers extension resources control policies. In: 26th {USENIX} Security Symposium ({USENIX} Security 2017), pp. 679–694 (2017)
52. Shan, H., Wang, Q., Pu, C.: Tail attacks on web applications. In: Proceedings of the 2017 ACM SIGSAC Conference on Computer and Communications Security, pp. 1725–1739. ACM (2017)
53. Silver, D., Jana, S., Boneh, D., Chen, E., Jackson, C.: Password managers: attacks and defenses. In: 23rd {USENIX} Security Symposium ({USENIX} Security 2014), pp. 449–464 (2014)
54. Song, Y., Cai, Z., Zhang, Z.L.: Multi-touch authentication using hand geometry and behavioral information. In: 2017 IEEE Symposium on Security and Privacy (SP), pp. 357–372. IEEE (2017)
55. Soska, K., Christin, N.: Automatically detecting vulnerable websites before they turn malicious. In: 23rd {USENIX} Security Symposium ({USENIX} Security 2014), pp. 625–640 (2014)
56. Su, Y., Genkin, D., Ranasinghe, D., Yarom, Y.: {USB} snooping made easy: crosstalk leakage attacks on {USB} hubs. In: 26th {USENIX} Security Symposium ({USENIX} Security 2017), pp. 1145–1161 (2017)
57. Tajalizadehkhoob, S., et al.: Herding vulnerable cats: a statistical approach to disentangle joint responsibility for web security in shared hosting. In: Proceedings of the 2017 ACM SIGSAC Conference on Computer and Communications Security, pp. 553–567. ACM (2017)
58. Tatlı, E.I.: Cracking more password hashes with patterns. IEEE Trans. Inf. Forensics Secur. **10**(8), 1656–1665 (2015)
59. Tian, D.J., Bates, A., Butler, K.R., Rangaswami, R.: ProvUSB: Block-level provenance-based data protection for USB storage devices. In: Proceedings of the 2016 ACM SIGSAC Conference on Computer and Communications Security, pp. 242–253. ACM (2016)
60. Valenta, M., Sandner, P.: Comparison of ethereum, hyperledger fabric and corda. [ebook] Frankfurt School, Blockchain Center (2017)
61. Vanhoef, M., Piessens, F.: Key reinstallation attacks: forcing nonce reuse in WPA2. In: Proceedings of the 2017 ACM SIGSAC Conference on Computer and Communications Security, pp. 1313–1328. ACM (2017)
62. Varadarajan, V., Zhang, Y., Ristenpart, T., Swift, M.: A placement vulnerability study in multi-tenant public clouds. In: 24th {USENIX} Security Symposium ({USENIX} Security 2015), pp. 913–928 (2015)
63. Vissers, T., Barron, T., Van Goethem, T., Joosen, W., Nikiforakis, N.: The wolf of name street: hijacking domains through their nameservers. In: Proceedings of the 2017 ACM SIGSAC Conference on Computer and Communications Security, pp. 957–970. ACM (2017)
64. Wash, R., Rader, E., Berman, R., Wellmer, Z.: Understanding password choices: how frequently entered passwords are re-used across websites. In: Twelfth Symposium on Usable Privacy and Security ({SOUPS} 2016), pp. 175–188 (2016)
65. Wood, G.: Ethereum: a secure decentralised generalised transaction ledger. Ethereum Proj. Yellow Pap. **151**, 1–32 (2014)

66. Wressnegger, C., Yamaguchi, F., Maier, A., Rieck, K.: Twice the bits, twice the trouble: vulnerabilities induced by migrating to 64-bit platforms. In: Proceedings of the 2016 ACM SIGSAC Conference on Computer and Communications Security, pp. 541–552. ACM (2016)
67. Xiao, Y., Li, M., Chen, S., Zhang, Y.: Stacco: differentially analyzing side-channel traces for detecting SSL/TLS vulnerabilities in secure enclaves. In: Proceedings of the 2017 ACM SIGSAC Conference on Computer and Communications Security, pp. 859–874. ACM (2017)
68. Yli-Huumo, J., Ko, D., Choi, S., Park, S., Smolander, K.: Where is current research on blockchain technology? A systematic review. PloS ONE 11(10), e0163477 (2016)
69. Zhang, Y., Juels, A., Reiter, M.K., Ristenpart, T.: Cross-tenant side-channel attacks in PaaS clouds. In: Proceedings of the 2014 ACM SIGSAC Conference on Computer and Communications Security, pp. 990–1003. ACM (2014)
70. Zuo, C., Zhao, Q., Lin, Z.: Authscope: towards automatic discovery of vulnerable authorizations in online services. In: Proceedings of the 2017 ACM SIGSAC Conference on Computer and Communications Security, pp. 799–813. ACM (2017)

Cancelable Palmprint Feature Generation Method Based on Minimum Signature

Jian Qiu, Hengjian Li[✉], and Xiyu Wang

School of Information Science and Engineering,
University of JiNan, Jinan 250022, China
ise_lihj@ujn.edu.cn

Abstract. With the emphasis on biometric privacy protection, this paper proposes a method for generating cancelable palmprint features based on minimum signature. Firstly, orthogonal features of ROI of palmprint are extracted. In order to realize a layer of security and cancelability of palmprint features, a chaotic matrix is randomly generated as the key, and then xor orthogonal features form the initial feature matrix. Then the signature matrix is generated by generating hash function randomly. The initial value of signature matrix is infinite. The initial feature matrix is scanned and calculated by using the generated hash function, and the larger value in the original signature matrix is replaced by the minimum value to form a final signature matrix as a cancelable palmprint feature stored in the database. Experiments and theoretical analysis prove that the scheme can maintain high recognition performance and effectively protect the privacy of palmprint.

Keywords: Cancelable palmprint · Privacy protection · Minimum signature

1 Introduction

With the development of science and technology, biometric recognition has been widely used. Biometrics technology overcomes the shortcomings of traditional identity authentication that are easy to be lost and forgotten. Its advantages such as uniqueness, convenience, stability, and high recognition rate have been rapidly popularized and applied. However, biometrics are unique. Once they are stolen, they will be stolen permanently and cannot be used for identity authentication. Therefore, biometric variability must be given to ensure its practicability.

Biometric template protection methods are mainly divided into three categories: biometric encryption protection method, biometric template transformation method and hybrid protection method [1]. Ratha [2] first proposed the concept of revocable biometrics. The final cancelable biometric is formed by Cartesian and polar coordinate transformation of the fingerprint minutiae points. However, it was later proved that the method was not safe enough to guarantee the safety of the original biometric features [3]. In the encryption protection method, Jules [4] proposed a Fuzzy Commitment method. In this scheme, error correction code technology is used to deal with the differences between intra classes. The most commonly used biometric template protection scheme is a transformation-based approach. In the research of iris template

R. Doss et al. (Eds.): FNSS 2019, CCIS 1113, pp. 118–127, 2019.
https://doi.org/10.1007/978-3-030-34353-8_8

protection, Umer [5] proposed a feature learning based protection scheme. In this scheme, two different tokens are used to modify the existing BioHashing technology. Dwivedi [6] used the scheme of decimal encoding and look-up table mapping to protect the iris template, but the recognition accuracy of the scheme depends on the extracted feature accuracy. In addition, there are a series of schemes using Bloom filter to protect iris template [7–9]. Although the Bloom filter scheme can effectively protect feature privacy, one problem is that space utilization is low. In terms of fingerprint template protection, Wang [10] proposed a blind identification scheme. This scheme is different from the traditional template protection scheme, but protects the frequency sample of binary string. In addition, he also proposed a cancelable fingerprint template protection scheme based on zoned minutia pairs [11]. The scheme is compact and computationally efficient and suitable for mobile devices. Alam [12] extended the 3-tuple quantization scheme of the polar grid to effectively protect the fingerprint template. Although there are so many biometric template protection schemes, these schemes can not be directly used for palmprint template protection. For palmprint template protection, Leng [1] proposed a row-alone and row-co-occurrence fuzzy vaults scheme. The scheme has high security and can resist several attacks. Li [13] also proposed a method for generating revocable palmprint features based on coupled nonlinear dynamic filters. In addition, Jin [14] et al. proposed an IOM-based template protection scheme in the biometric template protection scheme. This scheme is based on the minimum hash and locality sensitive hash scheme in the field of document retrieval. In this scheme, the hash function is used to replace the permuting in the original scheme, which makes the scheme simpler and safer. Based on this, this paper proposes a method for generating cancelable palmprint features based on minimum signature.

In this paper, we first use the security of chaotic matrix to generate initial feature matrix, which forms a layer of security protection. Then the irreversibility of hash function is used to scan the initial feature matrix to form a cancelable signature matrix. The minimum signature matrix forms another layer of security for the original palmprint feature. The program can effectively protect the security and privacy of palmprint biometrics, and can improve security and privacy while ensuring recognition rate.

This paper is organized as follows: In Sect. 2, orthogonal filter is introduced. In Sect. 3, minimum signature is introduced. The basic scheme is given in Sect. 4. In Sect. 5, experimental results and analysis are shown. In Sect. 6, the security is analyzed in detail. The conclusion is given in the last section.

2 The Theory of Orthogonal Filter

Gabor filters are widely used to extract palmprint features for authentication. There are many methods to extract palmprint features [15–17]. However, due to the wide frequency band of the Gabor filter, the palmprint structure cannot be effectively characterized. Orthogonal filter has narrow frequency band and is a smooth low resolution function in the contour direction [18]. In addition, the behavior of orthogonal filter in the orthogonal (singular) direction is similar to wavelet, which can better extract the palmprint feature information. The Anisotropic Filter (AF) filter can extract the

direction information of the palmprint feature very well. We can construct the orthogonal filter by AF. The expression of AF is as follows:

$$G(\mu, v) = (4\mu^2 - 2)\exp(-(\mu^2 + v^2))$$ (1)

The expressions for μ and v are as follows:

$$\begin{bmatrix} \mu \\ v \end{bmatrix} = \begin{bmatrix} 1/\alpha & 0 \\ 0 & 1/\beta \end{bmatrix} \begin{bmatrix} \cos\theta & \sin\theta \\ -\sin\theta & \cos\theta \end{bmatrix} \begin{bmatrix} x - x_0 \\ y - y_0 \end{bmatrix}$$ (2)

Where $[x_0, y_0]$ represents the center of the AF. Where θ represents the direction of the AF. In addition, α and β are scaling factors along the x-axis and the y-axis. Inspired by the original measures [19], the orthogonal filter is constructed as follows:

$$OF(\theta) = G(x, y, \theta) - G(x, y, \theta + \frac{\pi}{2})$$ (3)

3   The Theory of Minimum Signature

The minimum signature scheme is an improvement on the minimum hash scheme. In the original minimum hash scheme, the final hash eigenvalue is obtained by arranging. But it's obviously cumbersome when there's too many permutations. Therefore, the minimum signature scheme simulates the arrangement by using hash function, which makes the scheme more simple. First randomly generate n hash functions to generate the minimum signature matrix. The initial value of the minimum signature matrix $sig(i, c)$ is infinite. Where i is the generated i-th hash function and c is the c-th column of the initial feature matrix. Then the initial feature matrix is scanned by hash function. Finally, the maximum value in the signature matrix is replaced by the minimum value obtained by hashing. The specific implementation process is as follows:

(1) To iterate over all the row vector.
(2) For each column in the row, if the value is 0 don't do any action. If the value is 1, then calculate $h_i(r)$. Where i represents the $i-th$ hash function. r represents the $r-th$ row.
(3) The value $sig(i, c)$ in the signature matrix is compared with the hash value $h_i(r)$. If $sig(i, c) > h_i(r)$, then replace $sig(i, c)$ with $h_i(r)$ If $sig(i, c) < h_i(r)$, then the value in the signature matrix remains the same.

Here is an example shown in Fig. 1. As can be seen from Fig. 1, the initial feature matrix is $[10001; 01001; 01101]$. Hash function $h_1(x) = (2x + 1)$ mod 4. Hash function $h_2(x) = (3x + 1)$ mod 5. Initial signature matrix. Since there are only two hash functions in this example, the size of the signature matrix is 2×5, and the initial values are infinite. First look at the first row, the first column and the fifth column value is 1. Then calculate $h_1 = 3, h_2 = 4$. In this case, the signature matrix becomes $sig = \begin{bmatrix} 3 & \infty & \infty & \infty & 3 \\ 4 & \infty & \infty & \infty & 4 \end{bmatrix}$.

Then look at the second row, the second column and the fifth column value is 1. Then calculate $h_1 = 1$, $h_2 = 2$. In this case, the signature matrix becomes $sig = \begin{bmatrix} 3 & 1 & \infty & \infty & 1 \\ 4 & 2 & \infty & \infty & 2 \end{bmatrix}$. Then look at the third row, the second column, the third column, and the fifth column value is 1. Then calculate $h_1 = 3$, $h_2 = 0$. In this case, the signature matrix becomes $sig = \begin{bmatrix} 3 & 1 & 3 & \infty & 1 \\ 4 & 0 & 0 & \infty & 0 \end{bmatrix}$. The final signature matrix is obtained by looping through all the row vectors.

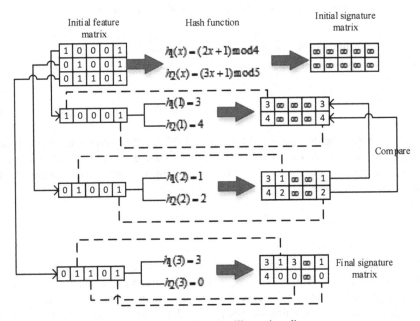

Fig. 1. Minimum signature illustration diagram.

4   Basic Scheme

In this section, we propose a cancelable palmprint template generation method based on minimum signature. This scheme realizes two-layer security protection for palmprint feature privacy. The general block diagram of this scheme is shown in Fig. 2 below.

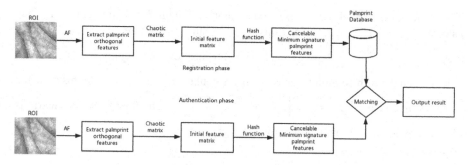

Fig. 2. The general block diagram of proposed method.

The specific implementation steps are shown below:

Step1: orthogonal filter is used to extract the orthogonal features of palmprint ROI. As follows:

$$F(x,y) = I(x,y) * OF(\theta) \tag{4}$$

Where $I(x,y)$ denotes a palm print image and "*" denotes a convolution operation.

Step2: In order to realize a layer of security and the cancelability of palmprint features, the initial feature matrix is formed by using the orthogonal features XOR chaotic matrix. The chaotic matrix is generated as follows:

$$O_{n+1} = \lambda O_n(1 - O_n) \tag{5}$$

Where $\lambda \in (0,4]$, $O_n \in [0,1]$.

Step3: Scanning the initial characteristic matrix by using the irreversibility of hash function. The final minimum hash signature matrix is stored in the database as the final cancelable palmprint feature. The specific implementation is shown in Sect. 3.

Step4: In the identity authentication, Jaccard similarity is used to distinguish. Jaccard formula as shown below:

$$J(A,B) = \frac{|A \cap B|}{|A \cup B|} = \frac{|A \cap B|}{|A| + |B| - |A \cap B|} \tag{6}$$

Where A and B represent sets satisfying certain conditions respectively.

5 Experiments and Analysis

5.1 Experimental Settings and Methods

In this section, to verify the minimum signature method authentication performance and security, we conducted a simulation experiment on the PolyU Palmprint Database. A total of 600 palm print images were included in the database. These palmprint images were collected to 100 different palms. Each palm was collected in two stage and three palm images were collected in each stage. Time interval was about two months. The 128×128 area of palmprint image center is extracted as the region of interest (ROI).

5.2 Minimum Signature Authentication

In the experiment, the palmprint feature size extracted by orthogonal filter is 32×96. In the experiment, the orthogonal features are divided into three $32 * 32$ blocks to form a signature matrix. In the experiment, two hash functions are used to form signature matrices: $h_1(x) = \mod(round(1000 \times O(0.8289, 0.2)), 32)$ and $h_2(x) = \mod(179 \times (x - 1) + 1, 32)$. Therefore, the size of the generated signature matrix is three 2×32 matrices. The final signature matrix formed here is the cancelable palmprint feature. A total of 179,700 matches were performed in the experiment, of which 1500 were intra-class matches and 178,200 were inter-class matches. To test the authentication performance of the scheme, the EER diagram of the scheme is shown in Fig. 2 below. As can be seen from Fig. 2, when the FAR is $10^{-2}\%$, the GAR is 97.14%. When the FAR is $10^{-1}\%$, the GAR is 98.16%. The EER obtained by using this scheme is 0.87%. From the Fig. 3 and the EER obtained, it can be concluded that the scheme can achieve better certification performance.

In order to reflect the performance of the scheme more intuitively, Fig. 3 shows the intra-class and inter-class matching score distribution curves.

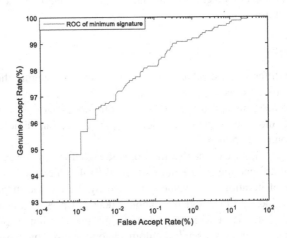

Fig. 3. The ROC curve of minimum signature.

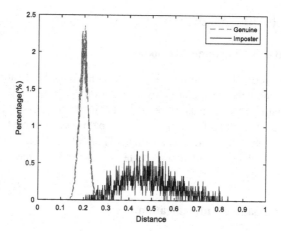

Fig. 4. Intra-class, inter-class score distribution curve.

As can be seen from Fig. 4, the intra-class matching scores are distributed around 0.197, and the inter-class matching scores are distributed around 0.485. As can be seen from the Fig. 3, it can be well distinguished between intra-class and inter-class.

6 Security Analysis

Biometric template protection should satisfy the following requirements: cancelability, diversity, irreversibility and accuracy [20]. Cancelability means that when the features in the database are threatened, new cancelable biometric features can be generated by changing the key and there is uncorrelation between these cancelable features. Diversity ensures that cancelable templates generated from the same original template can be applied to multiple different systems. Irreversibility means that when the biometrics in the database are stolen, the original biometrics cannot be recovered through the cancelable features.

6.1 Cancelability

We can generate different cancelable palmprint features by changing the initial values of the chaotic matrix. Logistic mapping is widely used because of its simple mathematical form, fast mapping speed and extremely complex dynamic behavior of chaotic system. Figures 5 and 6 show the bifurcation diagram of the one-dimensional Logistic map and its Lyapunov exponent.

From Figs. 5 and 6, we can see that the Logistic map does not always show chaotic state when the control parameters are in the interval $(0, 4]$. When $0 < \lambda \leq 3.56994$, after a certain number of iterations, the value generated by the chaotic mapping will converge to a fixed value, and the Logistic mapping does not present a chaotic state. But when $3.56994 < \lambda \leq 4$, the sequence generated through iteration is pseudo-random. Therefore, we can generate a pseudo-random chaotic matrix by reasonably setting

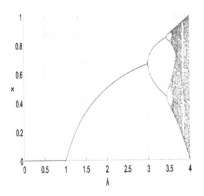
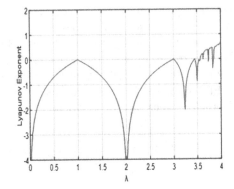

Fig. 5. Logistic chaotic map bifurcation diagram.   **Fig. 6.** Logistic chaotic map Lyapunov exponent.

parameters, and then generate a variety of cancelable palmprint features. Therefore, the cancelability and uncorrelation of palmprint features can be guaranteed by the pseudo-randomness of chaotic matrices.

6.2 Diversity

The diversity of biometrics can generate a variety of cancelable biometrics by changing the initial values of the chaotic matrix. Because in the minimum signature scheme, the initial feature matrix is obtained by orthogonal feature XOR chaotic matrix. Therefore, if the chaotic matrix is changed, the initial feature matrix is changed, and then the minimum signature matrix is changed. Therefore, a variety of cancelable features can be generated. Secondly, by changing the number of hash functions and their formulas, new and various revocable features can be generated. The size of the orthogonal feature extracted in the experiment of this scheme is 32×96. Then the orthogonal features are divided into three equal-sized blocks. Two hash functions are used in the experiment. Therefore, three 2*32 size signature matrices are generated as the final revocable palmprint feature. So by changing the number of hash functions, assuming that there are n hash functions then the resulting revocable palmprint feature is $3 \times n \times 32$. In addition, new cancelable features can be generated by changing the hash function formula. Therefore, the method meets the requirement of diversity.

6.3 Irreversibility and Accuracy

When cancelable palmprint features are stolen in the database, it is impossible for an attacker to recover the original palmprint features. First, the attacker does not know the formula for the hash function and the number of hash functions. Second, even if the attacker knows the specific hash function, because of the irreversibility of the hash function, the initial feature matrix cannot be generated from the cancelable features. Thirdly, even if the attacker gets the initial feature matrix, the pseudo-randomness of the chaotic matrix also guarantees the security of the original template. Based on the

above three points, the scheme meets the requirements of irreversibility. In addition, the key stolen cases also need to consider. we experimented with the recognition performance in the case of key theft, as shown in Fig. 5. As can be seen from Fig. 7, when the key is stolen, FAR is $10^{-2}\%$, the GAR is 96.05%. When the FAR is $10^{-1}\%$, the GAR is 98.09%. The EER obtained by using this scheme is 1.02%. From the Fig. 7 and the EER obtained, it can be concluded that it can still maintain good recognition performance when the key is stolen.

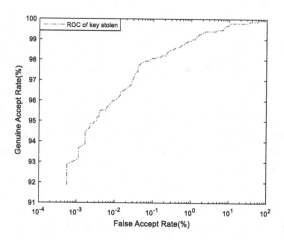

Fig. 7. The ROC curve of the key stolen.

7   Conclusion

In this paper, a scheme for generating cancelable palmprint features based on minimum signature is proposed. The chaotic matrix is used to XOR the extracted palmprint orthogonal features. The pseudo-randomness of the chaotic matrix implements a layer of security protection. Secondly, the initial feature matrix is scanned by the irreversibility of hash function. The final cancelable palmprint feature of the signature matrix implements another layer of security protection. Experiments show that the scheme can maintain good authentication performance. Moreover, through the security analysis, the scheme can effectively protect the privacy of palmprint.

References

1. Leng, L., Teoh, A.B.J.: Alignment-free row-co-occurrence cancelable palmprint Fuzzy Vault. Pattern Recogn. **48**(7), 2290–2303 (2015)
2. Ratha, N., Connell, J., Bolle, R.: Enhancing security and privacy in biometrics-based authentication systems. IBM Syst. J. **40**(3), 614–634 (2001)
3. Quan, F., Fei, S., Anni, C., Feifei, Z.: Cracking cancelable fingerprint template of Ratha. In: International Symposium on Computer Science & Computational Technology. IEEE (2008)

4. Juels, A., Wattenberg, M.: A Fuzzy commitment scheme. In: Proceeding of 6th ACM Conference on Computer and Communications Security, pp. 28–36 (1999)
5. Umer, S., Dhara, B.C., Chanda, B.: A novel cancelable iris recognition system based on feature learning techniques. Inf. Sci. **406–407**, 102–118 (2017)
6. Dwivedi, R., Dey, S., Singh, R.: A privacy-preserving cancelable iris template generation scheme using decimal encoding and look-up table mapping. Comput. Secur. **65**, 373–386 (2017)
7. Rathgeb, C., Breitinger, F., Busch, C.: Alignment-free cancelable iris biometric templates based on adaptive bloom filters. In: Proceedings of ICB, pp. 1–8 (2013)
8. Sadhya, D., Singh, S.K.: Providing robust security measures to Bloom filter based biometric template protection schemes. Comput. Secur. **67**, 59–72 (2017)
9. Bringer, J., Morel, C., Rathgeb, C.: Security analysis and improvement of some biometric protected templates based on Bloom filters. Image Vis. Comput. **58**, 239–253 (2017)
10. Wang, S., Hu, J.: A blind system identification approach to cancelable fingerprint templates. Pattern Recogn. **54**, 14–22 (2016)
11. Wang, S., Yang, W., Hu, J.: Design of alignment-free cancelable fingerprint templates with zoned minutia Pairs. Pattern Recogn. **66**, 295–301 (2017)
12. Alam, B., Jin, Z., Yap, W.S.: An alignment-free cancelable fingerprint template for bio-cryptosystems. J. Netw. Comput. Appl. **115**, 20–32 (2018)
13. Li, H., Zhang, J., Zhang, Z.: Generating cancelable palmprint templates via coupled nonlinear dynamic filters and multiple orientation palmcodes. Inf. Sci. **180**(20), 3876–3893 (2010)
14. Jin, Z., Lai, Y.L., Hwang, J.Y.: Ranking based locality sensitive hashing enabled cancelable biometrics: index-of-max hashing. IEEE Trans. Inf. Forensics Secur. **13**(2), 393–407 (2017)
15. Zhang, D., Zuo, W., Yue, F.: A comparative study of palmprint recognition algorithms. ACM Comput. Surv. **44**, 2–38 (2012)
16. Zhang, D., Kong, W.K., You, J., Wong, M.: Online palmprint identification. IEEE Trans. Pattern Anal. Mach. Intel. **25**, 1041–1050 (2003)
17. Fei, L., Xu, Y., Zhang, D.: Half-orientation extraction of palmprint features. Pattern Recogn. Lett. **69**, 35–41 (2016)
18. Li, H., Zhang, J., Wang, L.: Robust palmprint identification based on directional representations and compressed sensing. Multimed. Tools Appl. **70**(3), 2331–2345 (2014)
19. Chai, Z., Sun, Z., Méndez-Vázquez, H., He, R., Tan, T.: Gabor ordinal measures for face recognition. IEEE Trans. Inf. Forensics Secur. **9**(1), 14–26 (2014)
20. Rathgeb, C., Uhl, A.: A survey on biometric cryptosystems and cancelable biometrics. EURASIP J. Inf. Secur. **2011**(1), 3 (2011)

Thermal Distribution Change of Optical Fiber Terminal Based NG-PON2 Link Security Monitoring Technology

Xiaokai Ye and Xiaohan Sun[✉]

National Research Center for Optical Sensing/Communications Integrated
Networking, Southeast University, Nanjing 210096, China
{230179462,xhsun}@seu.edu.cn

Abstract. Physical fiber link security issues of the next generation of passive optical network (NG-PON2) have long been a concern. A fiber bending heat loss model have established, and proposed a method to monitor eavesdropping of transmission information in the fiber by detecting the thermal distribution of the output end of optical fiber links. Finally, the NG-PON2 link security monitoring scheme based on this method is given. The experimental results show that the proposed scheme can monitor the tapping in NG-PON2 fiber links.

Keywords: NG-PON2 · Security · Tapping · Eavesdropping · Thermal distribution

1 Introduction

NG-PON2 has become the most mainstream broadband fiber access method with the advantages of fiber to 5G (antenna feeder system), fiber-to-the-home and unique broadband networking [1, 2]. Because of the good compatibility with Ethernet as well as high bandwidth, long transmission distance and passivity between devices, NG-PON2 technology needs to work under high input optical power. At the same time, network security and reliability have been already listed as important features in the ITU.T-G.989 protocol [3, 4]. Once the upstream and downstream optical signals of the NG-PON2 are eavesdropping along optical fiber links, a large amount of data leakage is bound to occur. Therefore, it is necessary to explore techniques that can detect the eavesdropping at the physical layer of NG-PON2 and find a suitable scheme to ensure the security of optical fiber links.

At present, the traditional passive optical network is used to detect the tapping of the leaked optical signal, the optical power reduction ratio is used to determine whether eavesdropping occurs [5, 6], but the input power of the NG-PON2 is high, and because the cost induced by nonlinear in optical fiber is higher, especially with the fiber bending and eavesdropping technology becoming more sophisticated [7, 8], obviously the power of the stolen light is too small to be accurately judged. An invisible channel based on an optical code division multiple access (OCDMA) signal can enhance physical layer security, but still cannot detect tapping events [9]. Quantum Key Distribution (QKD) can detect any eavesdropping events through qubit security key

© Springer Nature Switzerland AG 2019
R. Doss et al. (Eds.): FNSS 2019, CCIS 1113, pp. 128–133, 2019.
https://doi.org/10.1007/978-3-030-34353-8_9

exchange, however, QKD consumes large wavelength resources and is not compatible with optical amplifiers, it is difficult to apply NG-PON2 architecture [10].

Therefore, it is extremely urgent to explore a novel real-time monitoring technology for tapping. So far, there is still a lack of means for detecting the physical quantity of the output end face of optical fiber links to acquire that the transmitted information in the fiber is eavesdropped. We propose that the thermal distribution of fiber end face changes greatly with the bending and torsion of the fiber at high fiber input power. Based on this effect, the possibility that the optical signal along the link is eavesdropped by attackers can be monitored; an NG-PON2 link security monitoring technology based on the thermal consumption of the fiber end terminal provides a guarantee for physical layer link security of the access network.

2 Principle

When the fiber is bent, the refractive index of the core layer (n_{core}) and the cladding layer (n_{cald}) will change, and the degree of change is closely related to the bending radius. It is known from the definition of the normalized frequency $V = ka\sqrt{n_{core}^2 - n_{clad}^2}$ that n_{core} increases when bending, then V becomes larger, and the amplitude of the optical pulse whose input energy is P changes, that is, P causes loss due to bending. Assuming that the bending coefficient is α, when tapping, the leakage energy ΔP per unit length of the fiber can be expressed as:

$$\Delta P = P \cdot \alpha = P \cdot \frac{2\alpha\kappa^2 e^{2\gamma a} \exp\left[-\frac{2}{3}\left(\gamma^3/\beta_g^2\right)R\right]}{e_v\sqrt{\pi\gamma R}V^2} \tag{1}$$

Where $\gamma = \left(\beta_g^2 - n_{cald}^2 k^2\right)^{1/2}$ and $\kappa = \left(n_{core}^2 k^2 - \beta_g^2\right)^{1/2}$ are related to n_{core} and n_{cald}, β_g is the propagation constant of the guided mode in the straight fiber, k is the free space propagation constant, a is the fiber radius, and R is the bending radius [11]. The ΔP is gradually converted into heat loss, which affects the thermal distribution state of fiber end face.

When fiber is bent, the thermal coupling effect is caused by mechanical motion, and the bending heat loss is generated. The bending and torsion of fiber produces stress and strain, which changes the bond length of O-Si-O and Si-O-Si, resulting in different degrees of displacement of the core and cladding SiO_2 network structure. Therefore, the fiber has micro-structural heat loss. It is represented by the radial displacement amount u_r and the thermodynamic coupled equation [12]:

$$\begin{cases} u_r = u = \dfrac{R^3 P_0}{(R+2a)^2 - R^3}\left[\dfrac{r}{3\lambda + 2u} + \dfrac{(R+2a)^3}{4ur^2}\right] \\ u_\theta = u_\varphi = 0 \end{cases} \tag{2}$$

$$\begin{cases} \rho_k c_k \frac{\partial T(r,t)}{\partial t} = \frac{\chi_k}{r^2} \frac{\partial}{\partial r} \left[r^2 \frac{\partial T(r,t)}{\partial t} \right] + Q \\ \frac{\partial}{\partial r} \left[\frac{1}{r} \frac{\partial}{\partial r} \left(\kappa^2 u_r \right) \right] = \alpha_k \frac{1+\gamma_k}{1-\gamma_k} \frac{\partial \theta(r,t)}{\partial r} \end{cases} \tag{3}$$

Where p_0 is the average pressure of the curved section, u is the elastic coefficient, α_k is the coefficient of thermal expansion, γ_k is the Poisson's ratio, ρ_k is the fiber density, c_k is the specific heat capacity of the fiber, r is the bending radius of a certain point ($R \leq r \leq R + 2a$), T is the temperature of the point; Q is the bending heat loss.

The tapping is monitored only by detecting the power, only the bending leakage energy ΔP is detected, and the bending heat loss Q that can characterize the tapping is almost ignored. The ΔP is converted into heat loss and detected together with the Q, that is, the heat consumption of the fiber terminal is detected, which can increase the accuracy of monitoring the tapping of the fiber links.

3   Experiment

3.1   Demonstration System

The schematic diagram of the NG-PON2 link security monitoring system based on the heat consumption of fiber terminal is shown in Fig. 1, which can be divided into a physical layer and a control layer. The physical layer includes couplers and thermal micro-scopes in addition to NG-PON2 basic devices; the control layer is composed of a memory, a processor, and a probe pulse laser.

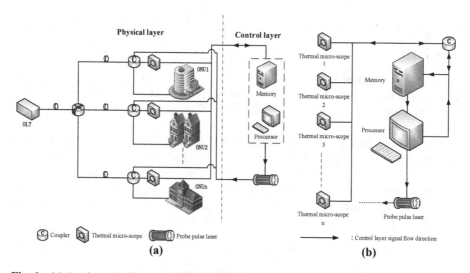

Fig. 1. (a) A schematic diagram of NG-PON2 link security monitoring system based on thermal distribution change of optical fiber terminal. (b) A diagram of control layer signal flow direction.

Initialize the calibration before using the system, collect the initial thermal distribution data of fiber end face through thermal micro-scopes, and transfer it to the memory for long-term storage until next re-initialization. Then, thermal micro-scopes monitor and collect the thermal distribution data of fiber end face, and transfers it to memory to compare with the data of the previous moment. If the data is error-free twice, then overlay storage, otherwise non-overlay storage. The processor processes the data in the memory in real time, and judges according to the thermal change (ΔT_i) of end points: i. all $\Delta T_i < T_1$, the fiber links has no heat loss, the system does not alarm; ii. all $T_1 < \Delta T_i < T_2$ (T_1 and T_2 vary depending on the environment), the fiber links normal heat loss, the system does not alarm; iii. any $\Delta T_i > T_2$, there is tapping in fiber links, the system issues an alarm. After occurrences 2 and 3, the processor controls the probe pulse laser to calibrate the system through different power detection pulses to restore the system to the initial state.

3.2 Experimental Results

We built the experimental system as shown in Fig. 2, and simulate an NG-PON2 link security monitoring based on the heat consumption of fiber terminal for one ONU user.

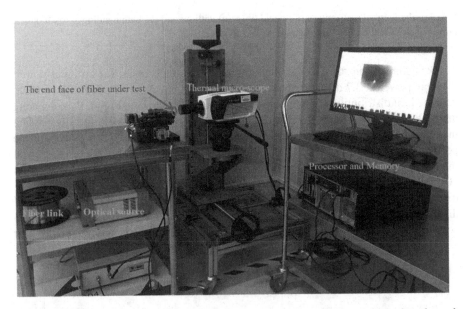

Fig. 2. A system diagram of NG-PON2 link security monitoring experimental based on thermal distribution change of optical fiber terminal.

In order to make the experimental phenomenon of tapping in optical fiber links more obvious, we observed two different cases: input power P = 3.7 dBm, without bending; input power P = 3.7 dBm, with bending. The thermal distributions are shown

in Fig. 3(a) and (b). Comparing Fig (a) and (b), it can be noticed that the tapping changes the overall fluctuation trend of thermal distribution on the end face of fiber links.

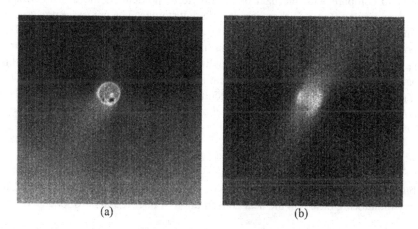

(a)   (b)

Fig. 3. The thermal distribution of optical fiber. (a) without and (b) with bending at input optical power of 3.7 dBm detected by thermal micro-scope.

We sampled the thermal distribution images for three cases: (a) no input power, (b) input power of 3.7 dBm without bending, (c) input power of 3.7 dBm with bending. And the heat distribution contour map of three cases, as shown in Fig. 4. By comparing the peaks (orange portion) of the three cases, it can be seen that when there is bending in optical links, the peak of the thermal distribution of fiber end face is lowered, and the peak-to-peak range is reduced. Then, comparing the green parts of the three cases, with increasing of input power, the thermal distribution of the fiber end face increases from 20.45–20.56 to 20.54–20.66, but when there is bending in the link, the thermal distribution is reduced to 20.53–20.63. From case (a) to (b) to (c), the median thermal distribution also increased from 20.48 to 20.60, and then decreased to 20.57. From the comparison results, it can be known that the fiber bending increases the link heat

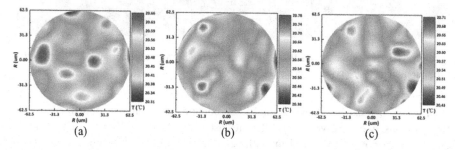

(a)   (b)   (c)

Fig. 4. The sampling contour plane of the thermal distribution. (a) no input power. (b) input power of 3.7 dBm without bending. (c) input power of 3.7 dBm with bending.

consumption and affects the thermal distribution of the fiber end face. Therefore, the NG-PON2 link security monitoring system based on the heat consumption of fiber end face can detect tapping event in optical fiber links and provide security for the physical layer fiber links of the access network.

4 Conclusion

This paper proposes a method for monitoring the eavesdropping of transmission information in the fiber by detecting the thermal distribution of the output end face of optical fiber links and the NG-PON2 link security monitoring scheme based on the method. The fiber bending heat loss model is given. The thermal distribution of the fiber end face is measured by thermal micro-scopes, and the change with the fiber tapping is obtained. The results show that the proposed method can monitor the tapping in physical layer fiber links.

References

1. Wey, J.S., Nesset, D., Valvo, M., Grobe, K., Roberts, H., Luo, Y.: Physical layer aspects of NG-PON2 standards—Part 1: Optical Link Design [Invited]. J. Opt. Commun. Netw. 8(1), 33 (2016)
2. Luo, Y., Roberts, H., Grobe, K., Valvo, M., Nesset, D., Asaka, K.: Physical Layer Aspects of NG-PON2 Standards—Part 2: System Design and Technology Feasibility [Invited]. J. Opt. Commun. Netw. 8(1), 43 (2016)
3. ITU-T: ITU-T Recommendation for 40-Gigabit-capable passive optical networks (NG-PON2)- General requirements. ITU-TRecommendatio G.989.1, pp. 1–22 (2013)
4. ITU-T: ITU-T Recommendation for 40-Gigabit-Capable Passive Optical Networks 2 (NG-PON2)- Physical Media Dependent (PMD) Layer Specification. ITU-T Recommendation G.989.2, pp. 1–108 (2014)
5. Fok, M.P., Wang, Z., Deng, Y., Prucnal, P.R.: Optical layer security in fiber-optic networks. IEEE Trans. Inf. Forensics Secur. 6(3), 725–736 (2011)
6. Kitayama, K.I., Sasaki, M., Araki, S., Tsubokawa, M., Tomita, A., Inoue, K.: Security in photonic networks: threats and security enhancement. J. Light. Technol. 29(21), 3210–3222 (2011)
7. Kiyokura, T., Uematsu, T., Manabe, T., Kawano, T., Hirota, H.: High-efficiency light injection and extraction using fiber bending. In: Optical Fiber Communications Conference & Exhibition. IEEE 2017, OFC, Los Angeles, USA (2017)
8. Uematsu, T., Hirota, H., Kawano, T., Kiyokura, T., Manabe, T.: Design of a temporary optical coupler using fiber bending for traffic monitoring. IEEE Photonics J. 9(6), 1–13 (2017)
9. Zhu, H., Wang, R., Pu, T., Fang, T., Xiang, P., Zheng, J.: Optical steganography of code-shift-keying OCDMA signal based on incoherent light source. IEEE Photonics J. 7(3), 1–7 (2015)
10. Yuan, C., Zhao, Y., Yu, W., Yu, X., Jie, Z.: Time-scheduled quantum key distribution (QKD) over WDM networks. J. Light. Technol. PP(99), 1 (2018)
11. Marcuse, D.: Curvature loss formula for optical fibers. Opt. Soc. Am. 66(3), 216–220 (1976)
12. Koldunov, M.F., Manenkov, A.A., Pokotilo, I.L.: On laser induced damage criterion. In: Laser-Induced Damage in Optical Materials, pp. 506–524 (1996)

Pseudonymization Approach in a Health IoT System to Strengthen Security and Privacy Results from OCARIoT Project

Sérgio Luís Ribeiro$^{(\boxtimes)}$ and Emilio Tissato Nakamura$^{(\boxtimes)}$

CPQD, Campinas, Brazil
{sribeiro,nakamura}@cpqd.com.br

Abstract. Regarding security and privacy in Internet of Things (IoT), especially in a digital health system, is necessary to guarantee that user rights are respected. This requires an approach that considers security-in-depth strategy established on risk-based results, actors, their privacy and the entire ecosystem, including the applications and platform. This paper presents an approach to strengthen the security and privacy aspects, using different security layers based on cryptographic, pseudonymization and anonymization technics to protect the processed, stored and transmitted data. The approach present at this paper was developed and applied in a digital health platform in the Project OCARIoT.

Keywords: Security · IoT security · Pseudonymization · Privacy · Digital health

1 Introduction

Security and privacy in Internet of Things (IoT), especially in a digital health system, it is important to understand the potential threats to that system, and add appropriate defenses accordingly, as the system is designed and architected. In the OCARIoT (Smart Childhood Obesity Caring Solution using IoT Potential) [1], it is not different, acceptances still depend on the certainty that the security and privacy rights are respected, in the whole system – device, applications and platform providers that requires a rigorous security-in-depth strategy.

It is why is important to design from the start with security and privacy in mind (Security by Design-SbD [2] and Privacy by Design-PbD [3] principles) because understanding how an attacker might be able to compromise a system helps to make sure appropriate mitigations are in place from the beginning.

This responsibility grows even more in Internet of Things (IoT) systems, since the fusion between the human, the digital and the physical exists. There are daily elements flowing digitally between heterogeneous components that are processing, transmitting and storing data.

Although security and privacy have a long history, it had been evolving since the very beginning when it surged in the information security science, where the main objective is to prevent sensitive information from reaching the wrong people, while making sure that the right people can in fact get it. However nowadays, with the advent

R. Doss et al. (Eds.): FNSS 2019, CCIS 1113, pp. 134–146, 2019.
https://doi.org/10.1007/978-3-030-34353-8_10

of new rules and laws related to privacy, such as GDPR [4], LGPD [5], Privacy Act [6] and others, the current evolving world, demands new and effective way to protect the privacy.

This paper presents an approach to strengthen the security and privacy, using security layers such as cryptographic, pseudonymization and anonymization elements to protect the processed (Data-In-Use, DIU), stored (Data-At-Rest, DAR) and transmitted (Data-In-Motion, DIM) data.

2   Anonymization and Pseudonymization Concepts

Anonymization and pseudonymization are two techniques that are recommended by the GDPR because they reduce risk and help in compliance with the data protection obligations. The main feature is to reduce the linking between the individual and the data, mainly after a data breach.

Anonymization is the permanent removal of any information that may serve as an identifier. Once a data set has been anonymized, it is impossible to identify individuals from it. Anonymizing data allows organizations to use the data for marketing and research, while protecting individuals from data exposure. However, since true anonymization is difficult to achieve, most businesses choose to use pseudonymization techniques [7].

Privacy protection is direct related to personal data, that GDPR defines as "any information relating to an identified or identifiable natural person ('data subject'). An identifiable natural person is one who can be identified, directly or indirectly, in particular by reference to an identifier such as a name, an identification number, location data, an online identifier or to one or more factors specific to the physical, physiological, genetic, mental, economic, cultural or social identity of that natural person" [8].

When done properly, anonymization can place data outside the scope of the GDPR. Some anonymization techniques, highlighted by the GDPR´s Article 29 Working Party (WP) issued Opinion 05/2014, includes [7]:

  i. **Noise Addition:** adding a level of imprecision to the original data. For example, a patient's weight shows a range of ± 7 kg., rather than a precise number.
 ii. **Substitution/Permutation:** replacing information with other values. For example, a patient's height of 100 cm might be stored as "blue."
iii. **Differential Privacy:** converting individual user data into something unidentifiable by bundling and blurring it in one way or another. Typically, differential privacy works by adding some noise to the data and the amount of noise added is a trade-off – adding more noise makes the data more anonymous, but it also makes the data less useful [9].
 iv. **Aggregation/K-Anonymity:** a "hiding in the crowd" concept where if each individual is part of a larger group, then any of the records in the group could correspond to a single person. For example, a data set might contain information about people in the São Paulo State instead of specifying a specific town, like Campinas.

Pseudonymization is defined in the GDPR as "the processing of personal data in such a manner that the personal data can no longer be attributed to a specific data subject without the use of additional information. Provided that such additional information is kept separately and is subject to technical and organizational measures to ensure that the personal data are not attributed to an identified or identifiable natural person." [8].

In other words, pseudonymization commonly refers to a de-identification method that removes or replaces direct identifiers (names, phone numbers, government-issued ID numbers, etc.) from a data set, but may leave in place data that could indirectly identify, that is considered a way to correlate various information to identify a person, often referred to as quasi-identifiers or indirect identifiers.

Applying such a method, and nothing else, might be called simple pseudonymization. Frequently, security and privacy controls designed to prevent the unauthorized re-identification of data are applied on top of simple pseudonymization to create strong pseudonymization [10].

3   Security and Privacy in OCARIoT Platform

This section presents some important issues related to security and privacy in order to protect the processed (Data-In-Use, DIU), stored (Data-At-Rest, DAR) and transmitted (Data-In-Motion, DIM) data by the OCARIoT Platform and to comply with privacy regulations like European GDPR (General Data Protection Regulation) [4] and Brazilian LGPD (*Lei Geral de Proteção de Dados*) [5].

The focus is on some particular topics: pseudonymization, data encryption and protection against side-channel attacks that are considered one of the most common attack in the privacy scenario. Other topics that complete the holistic view for the security and privacy in OCARIoT Platform, that are under development, especially the results from the risk assessment [11].

3.1   Identities and Accesses Premises

OCARIoT Platform – Is a Platform that deals with children information related to health and habits information that can be collect manually via Application, Web Dashboard or via IoT sensors that can be personal sensor, such as smart band or environmental sensor.

OCARIoT Platform is accessed by a set of different entities and needs an access control policy accordingly to the data, privacy requirements and the entity. Some premises are:

- OCARIoT Platform is accessed by the following entities:
  - Application: children.
  - Web Based Dashboard: parents, healthcare professionals, educators, and platform admin.
- Real children or natural children have a correspondent identification in the system (Child ID).

- OCARIoT Platform does not identify natural children. Real parents or natural parents have a correspondent identification in the system (Parents ID).
- OCARIoT Platform does not identify real parents.
- Healthcare professionals do not need to know the natural children.
- Data Privacy Officer (DPO) is the school representative that has the access to the natural children and parent's information.
- Platform Administrator (PA) is the entity that has access to the OCARIoT Platform and input data into the system. This can be done by terminal or script.
- PA takes data from DPO to include them into the system.
- DPO/PA need to link the natural child to his Child ID/pseudonym.
- DPO/PA need to link the parents to the correspondent child, in the natural and ID form.

3.2 Considerations

There are some considerations about the usage of Child ID and Parents ID:

- The child to access the app uses Child ID.
- Child ID is used by the healthcare professional to insert child data into the OCARIoT Platform.
- The parents to access the specific child data use parents ID.
- Healthcare professional uses Child ID to insert child related health data into the system.
- Healthcare professional does not need to know the Parents ID.
- There is a need to recover or reissue Child ID or Parents ID.
- Only DPO has knowledge about the natural child.
- DPO interacts with PA. PA interacts with the OCARIoT Platform using the information provided by DPO.

Regarding the use of Child ID by the healthcare professional, a process must define how he will take the knowledge about the Child ID. One possibility is to ask to the child to input his Child ID, and other are to ask for the Child ID to the child and select it directly from the system.

There are technological, usability and security issues related to each approach. In the first option, it is not necessary to the OCARIoT Platform to store the Child ID (only its correspondent hash (Sect. 2), what increases the security but interfering in the usability.

In the second approach, the healthcare professional can hear the Child ID and select it from the system menu, what make this user-friendlier, but on the other hand decreases security since a table of Child ID (instead of the hashed Child ID) must be stored in the OCARIoT Platform.

The same reason about technology, usability and security can be used to the use of groups of children. One option is to create groups by selecting the available Child IDs directly in the OCARIoT Platform interface. This approach has advantages in usability, but has to store the Child IDs into the system, what decreases security.

Other option is to creating groups without the OCARIoT Platform knowing the Child IDs, only their correspondent hashed versions. In this case, PA must input each Child ID or use a script to create groups, what decreases usability.

3.3   Initial Setup

This initial setup is related to the parents and children ID creation. There are four macro steps:

i. Data Privacy Officer (DPO)/Platform Administrator (PA) choose an ID for the child, Child ID.
ii. DPO/PA creates ID for the parents, Parents ID.
iii. DPO/PA links Child ID to Parents ID.
iv. DPO/PA links sensors to the Child ID.

As only DPO has the access to the natural children and parents, there is a need to DPO to pass the information to the PA to be inserted into the OCARIoT Platform.

This is a process outside from the OCARIoT Platform, and can be as simple as a spreadsheet or another external process.

3.4   Identity and Accesses

The basic components for OCARIoT Platform identity and accesses are in the Fig. 1.

The entities accessing the OCARIoT Platform represented in the figure are the child, Platform Administrator (PA), parents and healthcare professional. The Data Protection Officer (DPO) and school are other entities that do not access OCARIoT Platform. Besides that, the School and Home, are the entities where the Sensors are located, sending information (automatically) to the Platform

The main elements in the Fig. 1 are described as:

1. DPO has the children information that is not included in the Platform.
2. DPO/PA choose a generated Child ID for each child, e.g. BR1, BR2, ...
3. OCARIoT stores only the hashed Child ID.
4. DPO/PA creates ID for the parents (process to be defined).
5. DPO/PA links children ID to Parents ID.
6. DPO/PA links sensors to the Child ID.
7. The child to access the app uses Child ID.
8. Parents ID is used to access the specific child data.
9. Healthcare professional does not need to know the child or Parents ID.
10. There is a need to recover or reissue Parents ID.
11. Analytics is performed over personal data database.
12. Personal data database uses encryption.
13. Children database is not in the OCARIoT Platform.
14. Children access app using the Child ID.
15. Healthcare professional uses Child ID to insert child data into the Platform.

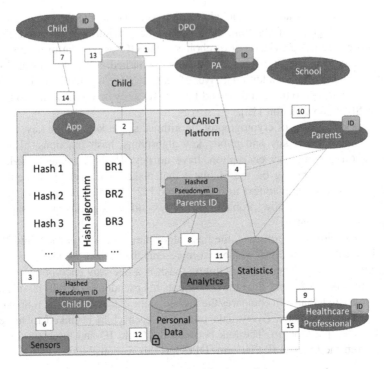

Fig. 1. Basic components for identity and access control.

4 Pseudonymization in OCARIoT Platform

OCARIoT Platform cannot use anonymization because of its nature to dealing with specific children data that requires historical, comprehensive and analytical data. Beyond that, data in OCARIoT Platform is dynamic, interacting with different entities such as healthcare professionals, parents, educators, technology provider and the children and also OCARIoT Platform needs methods to relinking the Child ID and Parents ID to the real or natural user.

The pseudonyms in the OCARIoT Platform are a set of characters that represents a natural user, like BR1, BR2 or as 12. The child and parents use these pseudonyms as an identification or login to access the platform. The basic authentication method is the password.

Considering that a security incident can leak the database, including the identifications, an additional layer of security is to don´t store the IDs directly in the OCARIoT Platform. In this case, what is stored in the platform is the hashed pseudonym. An issue about this additional security is the affected usability, since there are no lists of IDs provided to the user, for instance. Instead of that, the user needs to input his ID, just like traditional login methods.

In the case of healthcare professional inserting a child data into the system, or in the case of creating a group of children, the IDs must be previously known and inserted, and not chosen from a provided list of choice in the OCARIoT Platform interface.

What is stored in the OCARIoT Platform is the hashed pseudonym instead of the pseudonym. This creates an additional layer of security since a potential leakage doesn´t provide direct access to the Child ID and Parents ID, only to a hash with 256 bytes as a result of SHA-256 function [12].

There are three types of cryptography algorithms: secret key, public key, and hash functions. Unlike secret key and public key algorithms, hash functions, also called message digests or one-way encryption, have no key. Hash is a fixed-length value resulting from a computing on the plaintext that makes it impossible for either the contents or length of the plaintext to be recovered [12]. Hash algorithms are effective because of the extremely low probability that two different plaintext messages will yield the same hash value.

In the OCARIoT Platform case, the hash algorithm applies over the Child ID and Parents ID, creating a 256 bytes value that represents the hashed pseudonyms. As it is not possible to revert a hash to the plaintext (hash is a one-way encryption), the hash algorithm must be used every time the user inputs his pseudonym (Child ID or Parents ID). The OCARIoT Platform compares the hashed value calculated at that time with the stored hashed pseudonyms. Every data related to each child is linked to his correspondent hashed pseudonym. The link between the Child ID and the natural child is outbound from the OCARIoT Platform. DPO has this responsibility

4.1   Pseudonymization Method for Child ID

DPO needs to know the Child ID in order to link it to the child, sensors and parents. This process needs to be defined, since it includes an external method for the DPO to perform the links and the OCARIoT Platform that does not know who the child is. The PA configures the linking.

Healthcare professional needs to know the Child ID to insert child data into the OCARIoT Platform. Once the Child ID is inputted, the OCARIoT Platform performs the hash function to generate the hashed Child ID that is compared with the stored hashed Child ID.

Figure 2 shows the pseudonymization method for Child ID. Child ID is a code generated by the system (e.g. BR1, BR2, BR3, ... or something randomized) that is chosen by the DPO/PA linking it to the natural child. This is a pseudonym know by the child and by DPO and will be used by the entities to interact with the OCARIoT Platform.

To add an additional layer of security, the pseudonym will be stored in the OCARIoT Platform as a hash (hashed Child ID). A SHA-256 hash algorithm [12] applied in the original Child ID to generate the hashed Child ID.

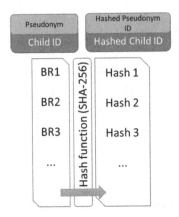

Fig. 2. Pseudonymization method for Child ID.

A hashed Child ID is the only information stored inside OCARIoT Platform (related to the Child ID). Every operation that uses the Child ID (e.g. BR1, etc.) performs a hash operation to compare the identity.

The authentication method for the child to access the application is password. The OCARIoT Platform using a similar approach for the identity protects the password, but using salt to increasing security, Fig. 3 shows the method to secure the passwords. Using salt, although the same password had been chosen by two different children (BR1 and BR2), they generate different hashes to be stored.

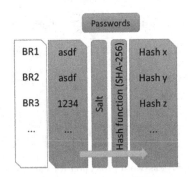

Fig. 3. Password protection for Child ID.

4.2 Pseudonymization Method for Parents ID

The pseudonymization method for Parents ID uses the same techniques from Child ID. The Parents ID is used by the parents to access the OCARIoT dashboard to access the information about their children. Besides the parents themselves, only the DPO needs to know the linking between the Child ID and the Parents ID.

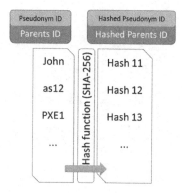

Fig. 4. Pseudonymization method for Parent ID.

The parent's pseudonym, Parents ID, can be chosen by the parents (easier to guess) or created by the OCARIoT Platform (more difficult to remember). A SHA-256 hash algorithm [12] is applied to the pseudonym to generate a hashed Parents ID. This hashed version of Parents ID is the only information stored inside OCARIoT Platform (related to the Parents ID) in order to provide an additional layer of security. Every operation that uses the Parents ID performs a hash operation to compare the identity (Fig. 4).

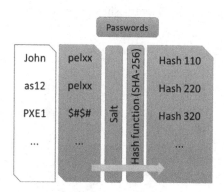

Fig. 5. Password protection for Parents ID.

Authentication method used by the parents to access the OCARIoT dashboard is based on the Parents ID and password Fig. 5. Each parent' correspondent passwords is stored in a salted hash format.

4.3 Data Encryption in OCARIoT Platform

The OCARIoT Platform architecture is based on microservices, which implies data distribution. The data encryption is not a requirement to all databases. There are the

GDPR and GDPL compliance requirements and the results of a risk assessment that direct the adequate security controls implementation by OCARIoT Platform. The risk assessment is under development and is an OCARIoT Project result.

Data encryption is a necessary additional layer of security that applies to all personal data. With encryption, a security incident like a leakage does not represent a direct compromise of children personal data confidentiality. According to the GDPR, personal data is "any information relating to an identified or identifiable natural person ('data subject'). An identifiable natural person is one who can be identified, directly or indirectly, in particular by reference to an identifier such as a name, an identification number, location data, an online identifier or to one or more factors specific to the physical, physiological, genetic, mental, economic, cultural or social identity of that natural person" [8].

OCARIoT Platform process, transmit or store different types of information. Some of them are considered personal data, such as questionnaires, notification, recommendation, reports, prescriptions, manually inputted personal data (like weight, height, restrictions), sensors data and logs. Others are not considered personal data, such as the environmental data, credentials or commands.

Data encryption applies to Data-In-Motion (DIM) and for Data-At-Rest (DAR). Identifying every points of attack in the OCARIoT Platform and understanding the interactions between different components is important to define the right security strategy. This is a result of risk assessment, which calculates the probability of threat agent exploring a set of vulnerabilities in one or more OCARIoT Platform components, turning a threat into a security incident, what causes impacts.

An application-level protocol like Transport Layer Security (TLS) [13] is suitable to protect DIM between internal components and between OCARIoT and users.

There are two possibilities to provide DAR encryption: in the application-level or in the database-level. In the application-level, cryptographic functions are used directly by the application in the server-side. In the database-level, there is the possibility to use a native encryption function in the database system or use third-part cryptography components integrated to the database system.

Symmetric cryptography algorithms are best suitable for DAR encryption since they are faster compared to asymmetric cryptography and a large range of computer processors supporting hardware-based cryptography. Advanced Encryption Standard (AES) is the standard algorithm for encryption [14].

Using symmetric cryptography has a challenge related to the key management. Important aspects need to be addressed like [15]: (i) the generation, changing and destruction of encryption keys; (ii) how and where the encryption keys are stored; (iii) how the keys are protected.

5 Protection Against Side-Channel Attacks

Side-channel attacks are a class of physical attacks in which an adversary tries to exploit physical information leakages such as timing information, power consumption, or electromagnetic radiation. Since they are non-invasive, passive and can generally be performed using relatively cheap equipment, they pose a serious threat to the security

of most cryptographic hardware devices. Such devices range from personal computers to small embedded devices such as smart cards and RFIDs (radio frequency identification devices [16].

In the OCARIoT Platform context, an attack flow can start in the children smart sensors, OCARIoT app or OCARIoT dashboard, alongside with the OCARIoT server-side services. An exploited vulnerability in any of these elements can lead to unauthorized access to information and leakage, modification or destruction. Proper OCARIoT Platform protection must prevent direct attacks against its components.

The protection against side-channel attacks in OCARIoT is more related to the sensors, mobile devices and hardware. In the software perspective, an appropriate memory management, especially for cleaning, is important. Processing OCARIoT information could lead to some timing, power or electromagnetic attacks.

The strategy for OCARIoT Platform is to reduce the attack surface by minimizing the use of personal information. Pseudonymization difficult the linking to the natural child and the stored information encryption protects against the data leakage.

To protect the OCARIoT Platform from attacks against, for instance, mobile device operational system, there are security techniques to run OCARIoT app in containers and use secure hardware modules to store sensitive information like cryptographic keys.

Security layers in OCARIoT Platform includes: (i) identification, authentication and authorization, (ii) data encryption, (iii) communications security and (iv) physical and asset security (by the cloud provider), this paper only present the data encryption security layer. Other security layers will be highlighted by the risk assessment [11] that is under development, including software development security, security engineering, security assessment and testing and security operations.

6   Future Work

In recent years, it is possible to notice a witnessed and exuberant wave of application possibilities for blockchain technology, since ensuring food safety and global self-sovereign digital identities, until decentralized virtual government management.

A blockchain technology is also appreciated, however aspects related to privacy, in some user case, should be considered and the approach presented at this paper can be used to strengthen security and privacy aspects.

In this way, security assessment and tests should be done in the project to have more experimental details and results to be present in other papers.

7   Conclusion

This paper presents an approach to strengthen the security and privacy, using different security layers such as cryptographic, pseudonymization and anonymization elements to protect the processed (Data-In-Use, DIU), stored (Data-At-Rest, DAR) and transmitted (Data-In-Motion, DIM) data, based on the results of the risk assessment.

This approach, considered that the OCARIoT Platform is in the design and development process by the following OCARIoT Project partners: Tecnalia Research & Innovation, Universidad Politecnica de Madrid, Centre for Research and Technology Hellas/Information Technologies Institute, Unparallel Innovation, Colegio Virgen de Europa, Servicio Madrileño de Salud, Ellinogermaniki Agogi, Universidade de Fortaleza, Instituto Atlântico, CPQD, Center for Strategic Health Technologies and Universidade Estadual do Ceará.

Besides the integration between different assets from the partners, the OCARIoT Platform also integrates third party components, sensors and infrastructure assets that were considered in the risk assessment.

Securing an Internet of Things (IoT) system, such as OCARIoT - due to the personal nature of the data collected - requires a rigorous security-in-depth strategy. In this way, to work with security and privacy in Internet of Things (IoT) is necessary to first look at the context where it will be used. Besides that, the best practices always suggest that is necessary to consider the whole system, starting from the context, then following all the layers, project, devices, applications, information data, network infrastructure and the platform. In the OCARIoT project, it is not different, acceptances still depend on the certainty that the security and privacy rights are respected, in the whole system – device, applications and platform providers.

Although GDPR does not refer to particular information such as: security standards, pseudonymization or anonymization technics. The use of these, provide a better security and privacy that will improve the acceptance by users regarding data security and privacy.

Consequently, an approach to strengthen security and privacy using different security layers, based on risk and implementing controls that encompass encryption, pseudonymization, and anonymization techniques to protect processed, stored, and transmitted data is required.

Acknowledgment. The authors acknowledge the financial support given to this work, under OCARIoT project, which received funding from the European Union's Horizon 2020 research and innovation programme under grant No. 731533 and the RNP under No. 3007. This paper reflects only the author's views and the Agencies are not responsible for any use that may be made of the information contained therein.

References

1. OCARIoT Project: Smart Childhood Obesity Caring Solution using IoT potential. https://ocariot.eu/. Accessed June 2019
2. OWASP: Security by design principles. https://www.owasp.org/index.php/Security_by_Design_Principles. Accessed July 2019
3. IAPP: Privacy by design 7 fundamentals principles. https://iapp.org/resources/article. Accessed July 2019
4. European Commission: 2018 reform of EU data protection rules. https://ec.europa.eu/commission/priorities/justice-andfundamental-rights/data-protection/2018-reform-eu-data-protection-rules_en. Accessed June 2019

5. Lei N°13.709: *Lei sobre a proteção de dados pessoais*. http://www.planalto.gov.br/ccivil_03/_Ato2015-2018/2018/L13709. Accessed June 2019
6. Privacy Act: Privacy Act of 1974. https://www.justice.gov/opcl/privacy-act-1974. Accessed July 2019
7. Perry, B.: Pseudonymization, Anonymization & GDPR. https://medium.com/@brperry/pseudonymization-anonymization-gdpr. Accessed June 2019
8. GDPR: General Data Protection Regulation. Art4. GDPR Definitions. https://gdpr-info.eu/art-4-gdpr/. Accessed June 2019
9. Valdez, A.C., Ziefle, M.: The users' perspective on the privacy-utility trade-offs in health recommender systems. Int. J. Hum.-Comput. Stud. **121**, 108–121 (2019)
10. Privacy Analytics: Comparing Pseudonymization and Anonymization Under the GDPR. http://www.privacy-analytics.com. Accessed July 2019
11. Ribeiro, S.L., Nakamura, E.T.: A privacy, security, safety, resilience and reliability focused risk assessment in a health IoT system - results from OCARIoT project. In: IEEE Global Internet of Things Summit (GIoTS), Arhus, Denmark, June 2019. ISBN 978-1-7281-2171-0
12. NIST: Hash Functions. Information Technology Laboratory. https://csrc.nist.gov/projects/hash-functions. Accessed May 2019
13. OpenSSL: Cryptography and SSL/TLS Toolkit. https://www.openssl.org/. Accessed June 2019
14. NIST: Block Cipher Techniques. Information Technology Laboratory. https://csrc.nist.gov/projects/block-cipher-techniques/bcm. Accessed June 2019
15. Business.Com: How to Select the Right Encryption Key Management Solution. https://www.business.com/articles/encryption-key-management-considerations/. Accessed June 2019
16. Verbauwhede, I.M.R.: Secure Integrated Circuits and Systems. Integrated Circuits and Systems. Springer, Boston (2010). https://doi.org/10.1007/978-0-387-71829-3

Security Analytics and Forensics

System Usage Profiling Metrics
for Notifications on Abnormal
User Behavior

Austin Sasko, Trevor Hillsgrove, Kanwalinderjit Gagneja[✉],
and Udita Katugampola

Florida Polytechnic University, Lakeland, FL 33805, USA
kgagneja@floridapoly.edu

Abstract. Due to greater media attention on security vulnerabilities more companies are investing in technical security. Even though technical measures do prevent technical attacks, these digital strongholds do nothing when considering social engineering vulnerabilities. Although security advancements can help prevent intrusion and illicit communications, they are powerless in stopping an already-authorized entity. In the cases of account-sharing, willful account turn-over ("fake technical support"), or remote desktop control, there are no measures in place with existing security solutions. Our proposed system addresses this weakness, through a user action profiler. As a user interacts with a system, a log of their usage is stored. This profile can then be used to detect action discrepancies. Our research focused on building an early profiler system and analyzed profile data evolution over time. This data will be eventually integrated into an anomaly detection system, which will be able to analyze incoming usage data and make a comparison to historical patterns. The intent is to effectively prevent any forms of account sharing or remote account owning, by requiring an alternate form of verification if a significant anomaly is detected.

Keywords: Application profiling · Account sharing · Social engineering · Vulnerabilities

1 Introduction

As more and more organizations increasingly use web and cloud resources for day-to-day business operations, greater opportunities for financial exploitation crop up. According to Morgan, cybercrime is expected to increase from $3 trillion to $6 trillion by 2021 [9]. Job openings in 2017 for skilled employees were "up from 1 million openings compared to 2018". Within this growing market area, social engineering is especially prominent and impactful. Approximately 84% of hackers surveyed at security conference DEFCON stated that they utilize some element of social engineering when carrying out their attacks [10]. Although technical vulnerabilities are important to secure, oftentimes the path of least resistance involves social manipulation. When considering industries that rely heavily on human resources and require data confidentiality, cyber protection becomes difficult to implement. Within the healthcare sphere, a survey revealed that "eighteen percent of healthcare employees are willing to sell confidential data to unauthorized

R. Doss et al. (Eds.): FNSS 2019, CCIS 1113, pp. 149–160, 2019.
https://doi.org/10.1007/978-3-030-34353-8_11

employees for as little as between $500 and $1,000" [8]. This survey does not even consider employees who may unintentionally reveal information or provide unauthorized access to a thought-to-be valid user. It becomes apparent that although technical security is necessary, some form of accountability is needed to help combat social manipulation, both willful (insider threat) and unknowing (trickery or carelessness).

1.1 Cost of Social Engineering

Social engineering exploits are particularly costly, as oftentimes their presence remains unnoticed for a significant amount of time. As they do not leave behind any obvious signs common in a more technical-based attack, social engineering is particularly damaging when attempting to clean up. Egan in his paper "Connecting the Dots: The Human Factor and the Cost of Cybercrime" explains that resolution time can take an average of 20 days to resolve phishing/social engineering attacks [11]. This time frame would vary based on the extent of the attack and how long it has progressed for. Outside of time resolution, monetary loss is significant for organizations who fall victim to social engineering. According to a report by Check Point Software Technologies, "social engineering attacks cost victims an average of $25,000–$100,000 per security incident" [12]. There is a high level of incentive for hackers in using social engineering techniques, and a great amount of risk for organizations should a breach succeed.

1.2 Difficulty of Prevention

Social engineering is seeing widespread usage and will continue to be used, due to the fact that it remains effective for obtaining information. Curtis Peterson, digital marketing manager for SmartFile, stated in an interview that "Social engineering is hard to prevent. That's the tough part (prevention)" [13]. With the widespread prevalence of social networking, lack of authenticity outside of a username, and human carelessness, a lot of attack vectors exist when forming social engineering exploit.

Social engineering prevention often requires direct interaction with every employee within an organization. Each individual must complete a comprehensive training procedure, which is a task that many smaller companies have difficulty in performing. Even with a dedicated training procedure, human oversight or mistakes may occur.

2   Related Work

Although social engineering is not necessarily preventable through traditional security methodologies, we see that training does play an important role. However, there are gaps if an organization relies only on education, which can be addressed in a more technical manner. This is the realm in which our proposed profiling system operates. By expanding upon existing authorization methods, it becomes possible to detect insider threat issues as they happen. Instead of monitoring network activity or scanning for virus signatures, each user is treated as a data-point that needs to be "scanned" every so often. This technical approach to a seemingly-nontechnical problem allows for

scalability and offers some means of fallback and quick response in case an attack slips past an employee's better judgment.

Insider threat is defined as "the potential for an individual who has or had authorized access to an organization's assets to use their access, either maliciously or unintentionally, to act in a way that could negatively affect the organization" [14]. At the present moment, companies do not have too many tools available at their disposal for the automated detection of any of the forms of insider threat. As a result, security researchers have begun to investigate this multi-faceted problem, beginning with a classification of insider threat. A recently published paper [5] details four categories of insider threat which could be improved upon with utility tools:

- Masquerading - user impersonation, account takeover
- Sabotage - Deliberate project or document destruction
- Information Theft - Document stealing
- Collusion - Illicit organizing.

2.1 Masquerading

An insider masquerading attack results when an attacker is able to pretend to be someone else. If a user finds an open workstation or commandeers a system through trickery, they are likely engaging in a masquerade attempt. Traits to monitor for detection and prevention can include biometrics, interface interaction, and keyboard/mouse input [5]. The system developed within this paper is meant to combat this form of insider threat. The following discussed systems also offer up useful ideas when addressing this problem.

Martín et al. in their paper, "Strokes of insight: User intent detection and kinematic compression of mouse cursor trails", involved mouse location tracing in an attempt to identify if the intent was involved in the given action or not [1]. By utilizing an implementation of kinematic compression, which states "a user in perfect control of his movements produces a minimum number of strokes in order to generate the intended trajectory" [1], the associated researchers were able to analyze an existing dataset and check for how and when movement related to action. The conclusion of this research culminated in a methodology suitable for distinguishing between intentional movements and unintentional movements. In terms of prevention of insider threat, a similar kinematic movement analysis was considered. By generating a profile over time of mouse actions, it may be possible to distinguish one entity from another. In case of a new user interacting with a system, hesitancy in mouse movement could be one indication of an abnormality, especially if the original user was known to be adept at interaction within the system.

Moriano et al. in their paper titled 'Insider threat event detection in user-system interactions' explain about insider threats with the intention of correlating abnormal actions with potential insider risk as engineers interacted with a variety of software components [7]. In order to form a baseline, a dataset of standard community action was recorded. This involved interaction with varying software tools utilized by engineers. Afterward, data from logs was analyzed before, during, and right after a major precipitating event took place. These precipitating events, like layoffs or company

changes, have been previously related to greater insider threat risk. By analyzing usage patterns of a user with different software tools, it was shown that there was a correlation between software abnormality interactions (actions outside of the average community operation) and precipitating events. As a result, it became possible to identify users who may have been adversely affected by the precipitating event. In the case of masquerading, this analysis can be useful in identifying compromised accounts. In the case of any major off-boarding process, it is possible to have a mix-up with account credentials. By looking for anomalies in account usage after major precipitating events, companies can identify and shut down potentially-misused accounts whose usage may have gone unnoticed otherwise.

2.2 Sabotage

Insider sabotage threats cover individuals who purposely "sabotage the integrity of their organization, often via sophisticated cyber attackers" [5]. These users are employees who deliberately attempt to bring about the downfall of a corporation for a variety of motives. Detection in this sphere is somewhat difficult, as the individual has insider knowledge and may attempt to bring harm without any extrinsic benefits towards themselves. In cases of information theft and masquerading, digital behavior anomalies are much easier to implement, as a noticeable recordable change occurs. However, with sabotage, it is possible that the actuator entered the organization with malicious intent from the beginning. An article 'A Human Factors Contribution to Countering Insider Threats: Practical Prospects from a Novel Approach to Warning & Avoiding' authored by Dr. Mils Hills and Anjali details the challenge of identifying a saboteur and proposes one possible approach [2]. Oftentimes, willful sabotage planning leaves random observable "pieces of information that may appear to be background noise at first glance," which can lead to clues when investigated further [2]. These traces, known as weak signals, are scattered across employees in the organization as people interact with others on a day-to- day basis. Although the aggregation of this material can pose arduous, it may have tangible benefits in preventing insider threat. The alternate method for detecting potential saboteurs proposed by Hills relies on sentinel events; seemingly random occurrences which have been shown historically to lead to a greater risk of sabotage. This method utilizes advancements in big data for collecting and classifying weak signals, HR notes, and any other relevant information. Over time this "Sentinel Event Database" would be able to identify possible forming sentinel events, leading to a reaction. Thus, although not directly measuring user actions as systems built to prevent information theft and masquerading threats, user profiling remains a strong candidate for early insider threat detection.

2.3 Information Theft

Information theft refers to the act of stealing confidential files from a system or organization or providing insider information to an unauthorized entity. In terms of danger to an organization, information theft is particularly problematic, especially if it involves patents or trade secrets. As this often involves network communication in cases with actual files, detection often involves monitoring and scanning abnormal transfers or

detecting abnormal logins [5]. In addition, an interesting research path resulted in theft prevention with intent based access control. Almehmadi and El-Khatib in their paper 'On the Possibility of Insider Threat Prevention Using Intent-Based Access Control (IBAC)' propose an intent-based access control [3] mechanism. One interesting detection mechanism introduced for combating malicious insider threat deals with measuring physiological brain signals. By running concealed information tests and measuring P300 wavelength response, the intent of a participant could be detected with a high degree of accuracy. This intent, in an IBAC, would be used along with standard authorization controls to grant or deny access to a user. In the testing environment, researchers utilized wavelength-detection devices with student participants. Each student was interrogated, and shown differing images of a burning lab, studying, or helping a person with their lab work. Whenever the P300 wavelength responded, the affiliated picture was noted. If the student reacted to the burning imagery, they would not be given access in an Intent-Based Access Control System. Alternatively, they were allowed access if there was no sign of malicious motivation. As applied to information theft, file access could be implemented with a similar detection system. If a user is both authorized and does not contain any malicious motivation, they will be granted access. However, if the malicious motivation is detected, further checks would be added in.

Mills et al. in their paper 'Predict insider threats using human behaviors' explained the concept of login monitoring and insider threat [4]. A more traditional research path for detecting information theft involved analysis on login attempts for a system resource. By recording and analyzing login and logoff events for a 9-month period, a large amount of profile data was generated for employees. After comparing habits to a baseline of activity, abnormal behavior was studied in more depth to check for relationships with suspicious activity. It was shown "that there is a relationship between users and their system login behaviors, such as total login events, suspicious login events, and suspicious login durations" [4]. In other words, by studying and comparing a baseline log activity with users, those who classified as threats were more likely to behave in an abnormal manner. Threats, in this case, were employees who had an unusual log history, which was mentioned as a precursor to information theft [5]. Once again, in cases involving insider threat, profiling an employee's behavior over time and detecting anomalies can be a particularly powerful tool, even in cases where a minimal technical attack was waged (e.g. masquerading or inadvertent information theft with social engineering techniques, or willful theft or sabotage by a disgruntled employee).

3 Problem Statement

It became increasingly noted while performing research that a majority of insider threat prediction techniques rely on accurate and detailed employee profiling over time. However, a majority of the investigated systems utilized previously-collected information, with modeling techniques applied retroactively. In all analyzed processes except for the Intent-Based Access Control (IBAC) system, data was pulled from organizations or from freely-available databases. Although useful for modeling approaches and for research, this does not prove too useful for active threat detection. The IBAC system [3] here differed, in that it actively collected wavelength data in real-time and made a

judgment almost instantaneously. This implementation as-is would not be easily scalable or implemented for current computer systems.

In order to address this problem of active user profiling over time, with the intent of preventing masquerading and account sharing, our profiling application will collect readily available average kinesthetic data (including mouse/keyboard holding and input per minute) per application used. This collected data, updated as time progresses, will eventually provide a baseline when checking for usage anomalies. In the case of a masquerading attempt, the system will be able to poll from its profiled information on a user-by-user basis and make a decision to allow the employee or challenge with a second form of authorization.

Therefore, the significance of our system is that it reliably collects the usage information for the purposes of further validating users. By operating as a background application capable of collecting a variety of behavioral input data, the proposed system both simultaneously creates a high-quality baseline for anomaly detection while remaining hidden on the system, thus offering opportunities for detection without alarming the masquerader. The user cannot claim for privacy as he has the responsibility to maintain the secrecy of the company data.

4   Methodology

For the initial prototype profiler, four metrics were chosen:

- Average keys typed per minute
- Average clicks per minute
- Average keystroke hold time
- Average mouse click hold time

These metrics were tracked with the following variables in our ProfileObj class:

```
private int timeExecuted; //how long the program has run for, total, in
seconds
private double kpm; //tracks average keys per minute
private double mcpm; //tracks clicks per minute
private int keysTyped; //holds total keypresses
private double aveIntKeyTime; //holds average key hold between key-
strokes
private int mousePress; //holds total mousepresses
private double aveIntMouseTime; //holds average mouse holds between
mouseclicks
```

As a user continues to interact with their system, each of these averages will update over time. By tracking total elapsed time and total keystrokes typed, each average will be as accurate as possible. For calculating the changing average upon updates, the following formulas were used:

$$kpm_{new} = (TM * kpm + keys)/(TM * timeMins) \qquad (1)$$

Where kpm = keys per minute, TM = total time elapsed, keys = keys pressed within the recorded interval, and timeMins = time for the added interval.

$$mcpm_{new} = (TM * mcpm + clicks)/(TM + timeMins) \qquad (2)$$

Where mcpm = mouse clicks per minute, TM = total time elapsed, clicks = clicks pressed within the recorded interval, and timeMins = time for the added interval.

$$ikt_{new} = (ikt * keys + held)/(keys + 1) \qquad (3)$$

Where ikt = average intermediary key hold time, keys = number of keys pressed and held = time that the most recent key has been held for.

$$imt_{new} = (imt * clicks + held) / (clicks + 1) \qquad (4)$$

Where imt = average intermediary mouse hold time, clicks = number of mouse clicks, and held = time that the most recent click has been held for.

For this study, approximately 11 h of interaction was recorded for a single user in order to establish a proof-of-concept, and to check if each metric showed promise as a profile facet.

In order to keep the data as realistic as possible, the profiler was only run when a subject was working on a programming task. As our prototype was restricted to only profiling programming-related applications like Terminal, Firefox Web Browser, and Atom Text Editor, and since we envisioned a use case for software engineers, it made sense to generate initial data pertaining to programming (Fig 1).

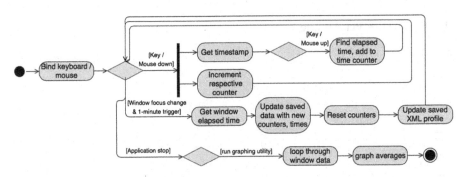

Fig. 1. Profiling activity diagram

5 Results and Analysis

For profile data collection, our profiler application was only run whenever programming tasks were performed by the user, to help retain consistency. In addition, three applications were chosen for profiling; Atom, Terminal, and Firefox. These three applications covered the minimum of what a programmer would need to utilize while working. In total 651 minute-long profile snapshots were collected throughout the data collection process, which contained all four metrics of interest broken-down by the application.

5.1   Mouse Clicks Per Minute

See Fig. 2.

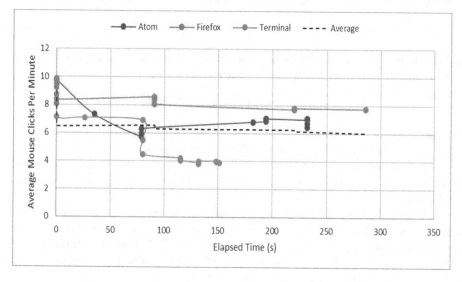

Fig. 2. Atom, Firefox, Terminal and average mouse clicks per minute

5.2   Keystrokes Per Minute

See Fig. 3.

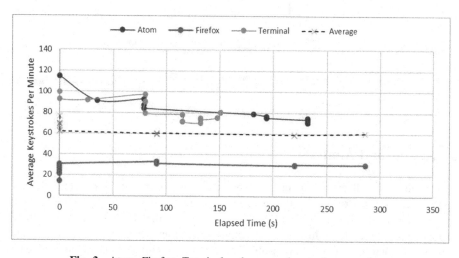

Fig. 3. Atom, Firefox, Terminal and average keystrokes per minute

5.3   Mouse Hold Time Per Click

See Fig. 4.

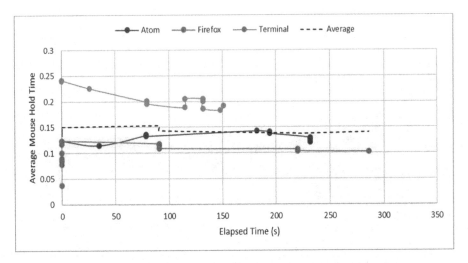

Fig. 4. Atom, Firefox, Terminal and average mouse hold time per click

5.4   Key Hold Time Per Click

See Fig. 5

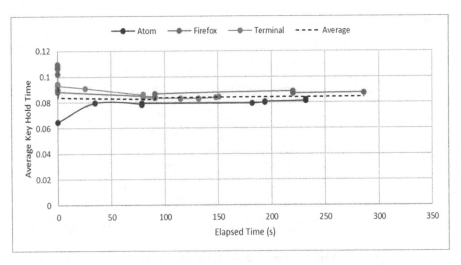

Fig. 5. Atom, Firefox, Terminal and average key hold time per click

5.5 Keystrokes Per Minute Scattered Plot

See Fig. 6.

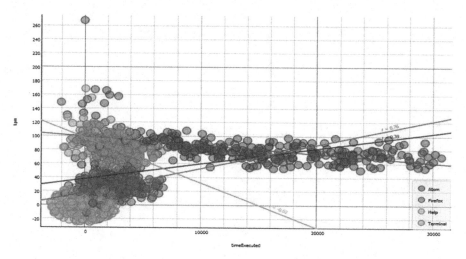

Fig. 6. Atom, Firefox, Terminal scattered plot for keystrokes per minute

5.6 ROC Plot

See Fig. 7.

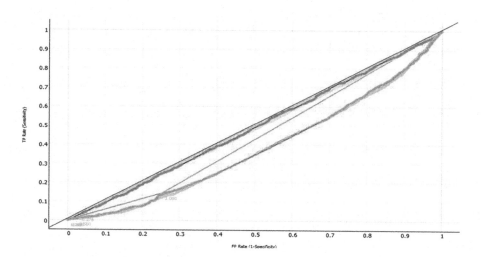

Fig. 7. Atom, Firefox, Terminal ROC plot

5.7 Analysis

From the above graphs, it is clear that each application does converge to an average, which is ideal when generating a user profile. The quickest metric observed was Firefox's mouse clicks per minute metric, with an approximate 391.553 s required to create a plateau on the graph. However, even with a quick convergence, we still see a significant amount of variability which should reduce as more data is collected. As shown in the Atom graphs, a significant amount of data over time does not completely remove variability; it instead reduces the overall variability from measurement to measurement. Atom's averaging out point was actually one of the longest recorded, due to the fact that collection within Atom made up the majority of the time breakdown.

From this, it can be deduced that all of the average metrics do eventually converge as expected, albeit with different variability depending on the application and metrics collected. The average "averaging out point" however gives us a good idea of when it is good to switch to the detection mechanism within our application. This critical time of $\sim 2,450$ s, or ~ 41 min, can be the golden time where our application makes the decision to switch over to an anomaly detection mode.

It is also observed that even after the estimated averaging point, many applications have a high amount of volatility. Firefox is especially subject to this, due to the fact that it is an application with access to many different websites. Terminal and Atom are somewhat restricted to one purpose, which is indicated by their relatively smooth graphs. However, browsers are by design multi-purpose, which presents a challenge when generating a profile solely on profiling alone.

One last interpretation of these results brought light to another potential identifying metric. Each application utilized within the 11 h spent profiling had differing amounts of interaction time, which led to differing elapsed time measurements. Atom, the text editor, by far had the most usage out of the three applications, with Firefox and Terminal roughly equaling each other in usage. In terms of user identification, this metric can be helpful, as it accurately profiles how a user spends his time on his machine.

6   Conclusion

According to the data collected during this prototype testing phase, profiling by utilizing the mouse and keyboard metrics appear to have promise. Each metric, given enough time, converged towards a somewhat-consistent value, which could be used to form a baseline for input checking. In addition, separating the data by application allowed for a cleaner, more useful data recording. One-task applications such as Terminal or Atom performed remarkably well, and data variability was not much of a factor. Firefox, understandably, had worse performance due to the wide range of functionality afforded by different web applications. However, from an overall standpoint, it appears this area of research has promising initial results and should be further developed and tested to test variance between different users.

References

1. Martín-Albo, D., Leiva, L., Huang, J., Plamondon, R.: Strokes of insight: user intent detection and kinematic compression of mouse cursor trails. Inf. Process. Manag. **52**(6), 989–1003 (2016)
2. Hills, M., Anjali, A.: A human factors contribution to countering insider threats: practical prospects from a novel approach to warning and avoiding. Secur. J. **30**(1), 142–152 (2017)
3. Almehmadi, A., El-Khatib, K.: On the possibility of insider threat prevention using intent-based access control (IBAC). IEEE Syst. J. **11**(2), 373–384 (2017)
4. Mills, J., Stuban, S., Dever, J.: Predict insider threats using human behaviors. IEEE Eng. Manag. Rev. **45**(1), 39–48 (2017)
5. Thing, V., Liau, Y., Divakaran, D., Ko, L.: Insider threat detection and its future directions. Int. J. Secur. Netw. **12**(3), 168 (2017)
6. Kloft, M., Laskov, P.: Online anomaly detection under adversarial impact. In: Teh, Y.W., Titterington, M. (eds.) Proceedings of the Thirteenth International Conference on Artificial Intelligence and Statistics, Series Proceedings of Machine Learning Research, vol. 9, pp. 405–412. Chia Laguna Resort, Sardinia (2010)
7. Moriano, P., Pendleton, J., Rich, S., Camp, L.J.: Insider threat event detection in user-system interactions. In: Proceedings of the 2017 International Workshop on Managing Insider Security Threats, pp. 1–12. ACM (2017)
8. Weak Healthcare Cybersecurity Employee Training Affects IT Security (2018). https://healthitsecurity.com/news/weak-healthcare-cybersecurity-employee-training-affects-it-security
9. Cybersecurity labor crunch to hit 3.5 million unfilled jobs by 2021 (2018). https://www.csoonline.com/article/3200024/cybersecurity-labor-crunch-to-hit-35-million-unfilled-jobs-by-2021.html
10. Fully 84 Percent of Hackers Leverage Social Engineering in Cyber Attacks (2018). https://www.esecurityplanet.com/hackers/fully-84-percent-of-hackers-leverage-social-engineering-in-attacks.html
11. Connecting the Dots: The Human Factor and the Cost of Cybercrime (2018). https://www.wombatsecurity.com/blog/phishing-social-engineering-the-human-factor-and-the-cost-of-cybercrime
12. Social engineering attacks costly for business (2018). https://www.csoonline.com/article/2129673/social-engineering-attacks-costly-for-business.html
13. Social Engineering: The Fastest Growing Threat to Business - Threat Sketch (2018). https://threatsketch.com/social-engineering-growing-threat-business/
14. CERT Definition of 'Insider Threat' – Updated (2018). https://insights.sei.cmu.edu/insider-threat/2017/03/cert-definition-of-insider-threat—updated.html

A New Lightweight Encryption Approach for Correlated Content Exchange in Internet of Things

Tasnime Omrani and Layth Sliman[(✉)]

EFREI Paris, Villejuif, France
layth.sliman@gmail.com

Abstract. So far, many cryptography systems adapted to Internet of Things (IoT) paradigm have been introduced. However, in these approaches, the constraint of images exchange has not been taken into consideration. In fact, image content is characterized by the correlation between its elements e.g. pixels. Correlation impacts negatively the security. Actually, using correlation, one can disclose data content once he/she has recognized a part of it. In this paper, we introduce a new cryptography system that takes into consideration correlation issue. Furthermore, our system allows reducing significantly memory use and time consumption, which are critical factor in IoT dedicated cryptosystems.

Keywords: Lightweight · Cryptography · Correlation · Confusion · Diffusion

1 Introduction

In IoT, objects are equipped with limited amounts of computational and storage capabilities. This is due to many physical, energy and cognitive constraints such as keeping a normal use of the object i.e. not to modify their essential characteristics and low energy consumption. Consequently, for each IoT object, a tradeoff between the supported level of security and the embedded computation capabilities should be reached.

One of the most resource consuming security functions is cryptography. Thus, to adapt to IoT context, the cryptosystems that are intended to be used in IoT applications should be optimized to lower processing and storage requirements while keeping a reasonable security with regard to the intended use of the secured object. To this end, many lightweight cryptosystems have been designed such as Present [1], MIPS [2], Rectangle [2], LED [4], Twine [5] Clefia [6], Midori [7], RoadRunner [8] and Simeck [9, 10]. However, neither of these systems has clearly considered the problem of correlated content exchange. In fact, correlated content such as images, multimedia and voice contents are involved in many IoT scenarios such as security camera, biometric authentication and autonomous vehicles, to cite few. Neglecting the removal of correlation of encrypted data content creates a major drawback in the existing lightweight cryptosystem.

R. Doss et al. (Eds.): FNSS 2019, CCIS 1113, pp. 161–171, 2019.
https://doi.org/10.1007/978-3-030-34353-8_12

Correlation is a statistical measure, which determines the capacity to predict a variable x by another one, y, using a linear relation [11]. The value of the correlation varies between +1 and −1:

- A value close to +1 indicates a strong correlation between the two variables i.e. the two variables vary in the same direction (increasing x will result in increasing y and vice versa).
- A value close to −1 indicate also a strong correlation but the variables vary in the opposite directions.
- A value close to 0 indicates no linear relation between the two variables.

In cryptography, the fact that two variables are highly correlated involves that the encrypted data are much easier to be disclosed by an attacker. In fact, the correlation can be used to predict some other features or parts of the intercepted data. For instance, images are characterized by the high correlation between there pixels. Thus, an attacker who succeeds to intercept a part of an image might be able to use the correlation to predict other parts of the image (e.g. the neighborhood of detected pixels). This requires that the encryption algorithm remove the correlation in data content.

As a result, to cope with IoT constraints, lightweight cryptography systems must tackle in the same time the problems of correlation in data content and the problem of resource consumption from the very beginning of their design.

To this end, in this paper, we describe a new lightweight cryptosystem sought to overcome the problem of correlation in data exchange, especially in IoT context.

In Sect. 2, called state of the art, we discuss the characteristics of some selected lightweight cryptosystems. The section contains an experimental part in which we show by simulation some characteristics of these cryptosystems, namely their resource consumption compared to the level of their output data entropy and correlation.

In Sects. 3 and 4, the new approach is implemented and analyzed. A comparison between our approach and Present is then achieved in Sect. 5. At the end of the paper we comment our results and describe our future work.

2   State of the Art

In this section, we carry out a quick analysis of a selection of popular and modern lightweight cryptosystems, namely Midori, Rectangle, Simeck and RoadRunner.

Midori encrypts a 64-bit or 128-bit message with a 128-bit key. The message consists of a matrix with 4 rows and 4 columns of nybbles in the case of 64 bits message, and with 4 rows and 4 columns of bytes in the case of 128 bits size message.

Midori uses SPN structure and involutive components. Using involutive components reduce memory use since the same functions are used for message encryption and decryption.

Table 1. Comparison of the studied cryptosystems

	Present	Rectangle	Midori	RoadRunner	Simeck
Key size	80 bits	80 bits	128	80	64
Block size	64	64	64	64	32
Memory for code	2.5 kb	2.2 kb	3.7 kb	2.7 kb	1.8 kb
Number of clock cycles	4 863 210	1 714 823	161 271	3 632 629	2 171 081
Entropy	7.997	7.951	7.997	7.994	7.934
Horizontal correlation	0.0031	0.0043	0.0303	0.0017	0.1425
Vertical correlation	0.0020	0.0066	0.0062	0.0117	0.341

RoadRunner encrypts an 80-bit message size. It is a Feistel-SPN architecture with 10 rounds with a 80 bit key or 12 rounds with 128-bit key. RoadRunner uses key whitening method to increase security.

Rectangle is a bit-sliced design. It encrypts a 64-bit message with a 64 or 128-bit key. The message is represented in form of a bits matrix with 4 rows and 16 columns. Rectangle uses SPN architecture with 25 rounds. The bit-sliced design makes the cryptosystem highly flexible for hardware implementation. Furthermore, due to columns parallelization in S-box function, the number of clock cycles is reduced.

Finally, Simeck uses only circular shifts and bitwise operations (XOR, AND) for round encryption and keySchedule function. This reduces memory and logic gates use.

In the Table 1 we show the different characteristics of the aforementioned cryptosystems. From the Table we can see that, except for RoadRunner, the value of correlations in the different studied cryptosystems should be optimized.

Although RoadRunner shows low correlations values, it tends to be resource greedy (time-wise).

In the following, we will introduce a new cryptosystem, called CRIoT (Correlation Removal for IoT) with the aim of tackling the problem of correlation, while keeping a good level of security and low resource consumption.

3 CRIoT Architecture

The goal of our proposal is to reduce the processing time and memory use while maintaining a low correlation of the data content along with a reasonable security level.

To improve the level of security we apply methods that provide a high confusion and diffusion levels while insuring a low transfer time. To remove the correlation, we use a high-level diffusion method. Finally, all the used techniques respect a low level of resource consumption.

To reduce processing time, we use Feistel architecture to ensure confusion [12]. The advantage of Feistel is that it needs to process only one-half of the block data during confusion. This significantly reduces the processing time. The resulted confusion is then diffused on all the block.

In the following, we explain the different parts of our architecture (Fig. 1).

CRIoT is a combination between SPN and Feistel. It consists of 11 rounds, the size of the block is 64 bits and the size of the key is 128 bits. Each block goes through the following steps:

- Apply a strong permutation method in order to remove the correlation. This method is inspired from CatMap [13]. We call this method "ECatMap" (Extended CatMap).
- Split the block into two sub-blocks (right and left).
- Apply an addition to a sub-key confusion function on the left sub-block.
- Apply a 4 bits S-box confusion function on the left sub-block.
- Aggregate the two sub-blocks.
- Apply a diffusion function on the entire block in order to diffuse the confusion all over the block so that the security level is improved.
- Re-split the block into two sub-blocks.
- Apply a non-linear and non-reversal function on the left sub-block.

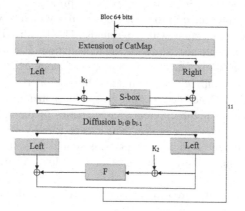

Fig. 1. Our cryptosystem

In the following, we explain the different components of our system.

3.1   Extension of CatMap

CatMap is a transformation that can be applied to an image. Once applied, the pixels of the image appear to be randomized. However, when the transformation is repeated enough times, the original image reappears. This facilitates the decryption and ensures a high level of diffusion in the cryptosystem. This function is applied to images by multiplying the image pixels coordinates by a matrix of the following form 1.

$$\begin{pmatrix} i' \\ j' \end{pmatrix} = \begin{pmatrix} 1 & a \\ b & a \times b + 1 \end{pmatrix} \times \begin{pmatrix} i \\ j \end{pmatrix} \bmod N \tag{1}$$

The problem of CatMap is that it can only be applied to images. In our system, to be able to deal with various type of data content, we represent data using vectors then we adapt CatMap in such a way that we can apply the transformation these vectors. To do so, we define a function $D = a \times b + 1$ with a $a \times b$ pair. This function takes as input a vector and returns another vector of the same characteristic. As we mentioned before, in CatMap, by repeating the function enough times, the initial data reappears. Similarly, in our case, and since we use a 64-bit size vector, by repeating the function 63 times, the initial vector is re-obtained.

3.2 S-Box

S-box is a substitution function which is usually represented as a table. The function takes a set of bits as input and transforms it into a set of output bits. It is worth mentioning that S-Box does not impose having the same number of bits at the input and at the output.

There are different approaches of S-box. The most popular is Rijndael S-box [14]. Rijndael S-box takes an input of 8-bits and produces an 8 bits output. In our case, we will use a 4 bits S-box i.e. an S-box that takes a 4-bit input and produces an output of 4 bits. This reduces the memory usage since a variable of 4 bits produces a substitution table of only 24 values. Furthermore, the involution property of the S-box makes S-box decryption function equivalent to encryption and hence reduce memory use. The used S-box is given Table 2.

Table 2. The used S-box

X	0	1	2	3	4	5	6	7	8	9	10	11	12	13	14	15
S(x)	C	A	D	3	E	B	F	7	8	9	1	5	0	2	4	6

3.3 The Non-Linear Function F

This is a non-linear function that applies the following formula to each pixel: $P = P^2 \oplus c$, with p is a pixel and c is a constant.

3.4 KeySchedule Function

Concerning the keySchedule function, the 128 bits key is divided in two equal parts; k_1 and k_2. k_1 is used for the part 1 of our cryptosystem and k_2 for the part 2. k_1 and k_2 are used for the rounds functions. They are the same for all rounds.

4 Experimental Results

For software implementation, the algorithm proposed in the previous section is implemented is implemented using VHDL description language and synthesized in Xilinx system design with SpartanIIE family of programmable IC and xc2s300e-6pq208 device.

In the following, we first introduce used parameters and measurement methods, then we show the simulation results.

4.1 Security Analysis

We are going to evaluate our proposal by applying some statistical tests such as Histogram, Entropy, Correlation, NPCR, and UACI. The description and the result of these tests are given in the flowing subsection.

Histogram

A histogram is a statistical curve, which indicates the distribution of pixels intensity values. In cryptography, the histogram of an encrypted image must be uniform to avoid the statistical attacks (Fig. 2).

As shown in the Fig. 3, we notice that by applying our proposal for Lena image, the initial image is not visible anymore. By comparing the histogram of the original image, Fig. 4, with the histogram of the encrypted image, Fig. 5, we can notice that the grey level is rather uniformly distributed.

Fig. 2. Original image

Fig. 3. Encrypted image

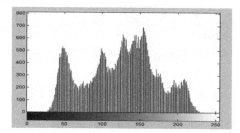

Fig. 4. Histogram of original image

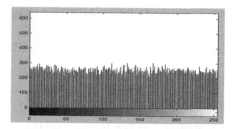

Fig. 5. Histogram of encrypted image

Entropy

Entropy [15] is a metric that assess the degree of uncertainty and randomness in the information. In image context, it measures the distribution of the gray value in an image. The more this metric is close to 8 (8 bits/pixel) the better the system is [16]. It's calculated using the formula 2,

$$E = \sum_{i=0}^{N}\left(Pi \times \log2\left(\frac{1}{Pi}\right)\right) \tag{2}$$

with Pi = the probability of appearance of symbol I and N equal to 256.

Entropy value of our cryptosystem is given in the following table. It's close to 7.9972, which is very close to 8 bits/pixel (Table 3).

Table 3. Entropy of original and encrypted image

	Original image	Encrypted image
Entropy	7.4003	7.997

Correlation Coefficient Analysis

Generally, images are characterized by the strongly correlated adjacent pixels.

An important requirement of an efficient cryptosystem is the generation of an encrypted image with a low adjacent pixels correlation. In fact, the more the adjacent pixels of the encrypted image are correlated, the easier an attacker can disclose its content. Thus, an efficient cryptosystem must have a correlation coefficient very close to 0. Correlation coefficient is given by the following equation [17]

$$C = \frac{\sum_{i-1}^{n}(x_i - \bar{x})(y_i - \bar{y})}{\sqrt{\left[\sum_{i-1}^{n}(x_i - \bar{x})^2\right]\left[\sum_{i-1}^{n}(y_i - \bar{y})^2\right]}} \tag{3}$$

To analyze the efficiency of our cryptosystem, the correlations between two adjacent pixels in horizontal and vertical dimensions are calculated. As shown in Table 4, the correlation coefficients for the original image are both close to 1. This shows that

the pixels are highly correlated. While for the encrypted image, the coefficients are close to 0.

Table 4. Correlation coefficient

	Original image	Encrypted image
Horizontal correlation	0.9256	0.0015
Vertical correlation	0.9606	−0.0012

NPCR (Number of Changing Pixel Rate):
NPCR measures the percentage of different pixels in two images. In cryptography, the optimal value of NPCR is equal to 99,6094 [18]. The value of NPCR is measured by the Eq. 4 [19].

$$D(i,j) = \begin{cases} 0, & if\ C^1(i,j) = C^2(i,j) \\ 1, & if\ C^1(i,j) \neq C^2(i,j) \end{cases}$$
$$NPCR: N(C^1, C^2) = \sum_{i,j} \frac{D(i,j)}{T} \times 100\% \tag{4}$$

The formula takes as input two copies of the same image with only 1 bit difference (one bit is altered so that we can check the impact of the modification in the encrypted image). In our case, the resulted NPCR is 99.60, very close to optimal value.

UACI (Unified Averaged Changed Intensity)
UACI measures the number of averaged changed intensity between ciphertext images. The optimal value of UACI is equal to 33.4635 [18]. The value of UACI is measured by Eq. 5 [19].

$$UACI: U(C^1, C^2) = \sum_{i,j} \frac{|C^1(i,j) - C^2(i,j)|}{F.T} \times 100\% \tag{5}$$

In our proposal the value of UACI of initial and encrypted images is 33.63 (very close to the optimal value).

4.2   Performance

Due to the limited computing resources embedded in IoT objects, before adopting any encryption system, it is essential to measure the memory use and encryption time it provides. In this section, we describe the measurement parameters such as ROM, RAM and processing time.

In our analysis we measured the processing time in term of number of clock cycles necessary for data encryption. This enables calculating the speed of the algorithm in seconds with respect to the performance of the device on which the encryption system is deployed.

Table 5. Comparison of the studied cryptosystems with our approach

	Present	Rectangle	Midori	RoadRunner	Simeck	CRIoT
Key size	80 bits	80 bits	128	80	64	128
Block size	64	64	64	64	32	64
Memory	2.5 kb	2.2 kb	3.7 kb	2.7 kb	1.8 kb	2.4 kb
Number of clock cycles	4 863 210	1 714 823	161 271	3 632 629	2 171 081	1 961 400
Entropy	7.997	7.951	7.997	7.994	7.934	7.997
Horizontal Correlation	0.0031	0.0043	0.0303	0.0017	0.1425	0.0015
Vertical Correlation	0.0020	0.0066	0.0062	0.0117	0.341	−0.0012

Calculating the amount of memory used by the system involves calculating the amount of used ROM and RAM by the algorithm to be executed for software implementation or the number of logic gates for hardware implementation. The amount of ROM used by the cryptosystem to be run matches to the encryption code size as well as the key size. Lightweight software implementations require up to 32 kb ROM [20]. While the used RAM matches to the machine code size, input data and the area of the memory used to store the intermediate data resulted from the execution of loops and recursive functions embedded in the code i.e. the stack. Lightweight cryptosystem software implementations require up to 8 kb RAM [20].

For hardware implementation, to consider that the cryptosystem as "lightweight", the number of used logic gates (GE) must not exceed 3000 GE [20].

Performance results of our cryptosystem are given in the following Table 6.

Table 6. Performance results

	Memory used (code)	GE	Numbre of clock cycles	Time
CRIoT	2.4 kb	2619.784	1 961 400	0.2 s

5 Comparison

In this section, we will proceed to a comparative analysis of the most known lightweight encryption algorithm, Present, as well as some of the recently proposed lightweight encryption systems, namely, Rectangle Midori, RoadRunner and Simeck.

The result of the comparison test is represented in Table 5.

We can conclude that, compared to the other cryptosystems, our cryptosystem addresses the best the correlation constraint due to its strong permutation method. Furthermore, our cryptosystem is characterized by a reduced number of clock cycles and reduced memory usage. This is due to the combination of Feistel and SPN architectures, and the choice of the components, respectively.

6 Conclusion

In this paper, we described CDIoT, a new lightweight block based encryption system that takes into consideration correlation constraint in data exchange.

The simulation and system performance assessment results show that the proposed encryption system leads to a better correlation coefficient compared to existing cryptosystems, while providing a higher level of security and a low memory usage and processing time. As a next step, we will proceed to the validation of our cryptosystem in a full-scale scenario by implementing it using embedded chips such as Smartcards and FPGA.

REFERENCES

1. Bogdanov, A., et al.: PRESENT: an ultra-lightweight block cipher. In: Paillier, P., Verbauwhede, I. (eds.) CHES 2007. LNCS, vol. 4727, pp. 450–466. Springer, Heidelberg (2007). https://doi.org/10.1007/978-3-540-74735-2_31
2. Izadi, M., Sadeghiyan, B., Sadeghian, S.S., Khanooki, H.A.: MIBS: a new lightweight block cipher. In: Garay, Juan A., Miyaji, A., Otsuka, A. (eds.) CANS 2009. LNCS, vol. 5888, pp. 334–348. Springer, Heidelberg (2009). https://doi.org/10.1007/978-3-642-10433-6_22
3. Zhang, W., Bao, Z., Lin, D., Rijmen, V., Bohan, Y., Ingrid, V.: RECTANGLE: a bit-slice ultra-lightweight block cipher suitable for multiple platforms. IACR Cryptol. ePrint Arch. **2014**, 84 (2014)
4. Guo, J., Peyrin, T., Poschmann, A., Robshaw, M.: The LED block cipher. In: Preneel, B., Takagi, T. (eds.) CHES 2011. LNCS, vol. 6917, pp. 326–341. Springer, Heidelberg (2011). https://doi.org/10.1007/978-3-642-23951-9_22
5. Suzaki, T., Minematsu, K., Morioka, S., Kobayashi, E.: TWINE: a lightweight block cipher for multiple platforms. In: Knudsen, L.R., Wu, H. (eds.) SAC 2012. LNCS, vol. 7707, pp. 339–354. Springer, Heidelberg (2013). https://doi.org/10.1007/978-3-642-35999-6_22
6. Shirai, T., Shibutani, K., Akishita, T., Moriai, S., Iwata, T.: The 128-bit blockcipher clefia (extended abstract). In: Biryukov, A. (ed.) FSE 2007. LNCS, vol. 4593, pp. 181–195. Springer, Heidelberg (2007). https://doi.org/10.1007/978-3-540-74619-5_12
7. Banik, S., Bogdanov, A., Isobe, T., Shibutani, K., Hiwatari, H., Akishita, T., Regazzoni, F.: Midori: a block cipher for low energy. In: Iwata, T., Cheon, J.H. (eds.) ASIACRYPT 2015. LNCS, vol. 9453, pp. 411–436. Springer, Heidelberg (2015). https://doi.org/10.1007/978-3-662-48800-3_17
8. Baysal, A., Şahin, S.: RoadRunneR: a small and fast bitslice block cipher for low cost 8-bit processors. In: Güneysu, T., Leander, G., Moradi, A. (eds.) LightSec 2015. LNCS, vol. 9542, pp. 58–76. Springer, Cham (2016). https://doi.org/10.1007/978-3-319-29078-2_4
9. Yang, G., Zhu, B., Suder, V., Aagaard, Mark D., Gong, G.: The Simeck family of lightweight block ciphers. In: Güneysu, T., Handschuh, H. (eds.) CHES 2015. LNCS, vol. 9293, pp. 307–329. Springer, Heidelberg (2015). https://doi.org/10.1007/978-3-662-48324-4_16
10. Kölbl, S., Roy, A.: A brief comparison of simon and simeck. IACR Cryptology ePrint Archive (2015)
11. Claude, G.: Initiation aux Methodes Statistiques en Sciences Sociales. Chapitre6 (1998). available at http://grasland.script.univ-paris-diderot.fr/STAT98/stat98_6/stat98_6.htm. Accessed 01 July 2016

12. Stéphane, J.: Protection cryptographique des bases de données: conception et cryptanalyse. PhD. Dissertation, Université Pierre et Marie Curie - Paris VI, (2012)
13. Soleymani, A., Nordin, M.J., Sundararajan, E.A.: Chaotic cryptosystem for image based on Henon and arnold cat map. Sci. World J. **2014**, 21 (2014)
14. Rijmen, V.: Efficient Implementation of the Rijndael S-box. J. Katholieke Universiteit Leuven (2000)
15. Skórski, M.: Shannon entropy versus renyi entropy from a cryptographic viewpoint. In: Groth, J. (ed.) IMACC 2015. LNCS, vol. 9496, pp. 257–274. Springer, Cham (2015). https://doi.org/10.1007/978-3-319-27239-9_16
16. Ahmad, M., Alsharari, H.D., Nizam, M.: Security improvement of an image encryption based on mPixel-chaotic-shuffle and pixel-chaotic-diffusion. Eur. J. Sci. Res. (2013)
17. Behnia, S., Akhshani, A., Ahadpour, S., Mahmodi, H., Akhavan, A.: A fast chaotic encryptions cheme based on piece wise nonlinear chaotic maps. Phys. Lett. A **366**, 391–396 (2007)
18. Jean, D.D.N.: Evaluation d'un algorithme de cryptage chaotique des images basé sur le modèle du perceptron, Report, Université de Ngoundéré (2012)
19. Wu, Y., Noonan, J.P., Agaian, S.: NPCR and UACI Randomness Tests for Image Encryption. J. Sel. Areas Telecommun. (JSAT) **1**, 31–38 (2011)
20. Manifavas, C., Hatzivasilis, G., Fysarakis, K., Rantos, K.: Lightweight cryptography for embedded systems – a comparative analysis. In: Garcia-Alfaro, J., Lioudakis, G., Cuppens-Boulahia, N., Foley, S., Fitzgerald, William M. (eds.) DPM/SETOP-2013. LNCS, vol. 8247, pp. 333–349. Springer, Heidelberg (2014). https://doi.org/10.1007/978-3-642-54568-9_21

Digital Forensics for Drone Data – Intelligent Clustering Using Self Organising Maps

Sri Harsha Mekala and Zubair Baig$^{(\boxtimes)}$

Deakin University, Geelong, VIC 3216, Australia
{shme,zubair.baig}@deakin.edu.au

Abstract. Drones or unmanned aerial vehicles (UAVs) have been rapidly adopted for a range of applications over the past decade. Considering their capabilities of information acquisition and surveillance for intelligence, and increasing access to the common public, have led to a rise in the threat associated with cyber crime associated with drones, in recent times. In order to mitigate the threats and to prevent cyber-crime, digital forensics on drone data is both critical as well as lacking in terms of efficacy studies. In this paper, we define a digital forensic methodology for analyzing drone data, and we propose a self organizing map (SOM)-based method for aiding such analysis. Experiments were conducted on two images obtained from the CFReDS project, namely, ArduPilot DIY Drone and DJI Phantom 4, with the purpose of producing admissible digital forensic evidence for the court of law, as part of a cyber-crime investigation. We also highlight the individual capabilities of several digital forensic tools based on experiments conducted on drone data.

Keywords: Drones or unmanned aerial vehicles (UAVs) · Threats and cyber crime · Digital forensic analysis · Digital forensic tools

1 Introduction

Drones, aka unmanned aerial vehicles, are increasingly being deployed to effect a large domain of applications including the delivery of packages to remote locations, taking pictures of critical events, to fire missiles in a battlefield, and in agriculture, for effective delivery of resources to hard-to-access locations [1, 3]. Drones are typically equipped with cameras to facilitate recording of pictures and videos, GPS for navigation, and a memory card to facilitate storage of recordings, for subsequent transfer to a processing/computing device [4]. Recent studies show that the increase of cyber threats and attacks has imposed a significant demand on digital forensic investigators and researchers alike for enhancing the digital techniques or methods adopted for accurately presenting criminal evidence in the court of law. With increasing sophistication of networking and computing infrastructures, construing evidence from large volumes of data available as a form of digital evidence, is becoming a hard task [2]. Therefore, the design of an efficient and effective digital forensic procedure, is of utmost significance to support criminal implication. A typical drone is a combination of the drone with air navigation capabilities, and a digital controller. Six common categories of drone applications include [2]:

© Springer Nature Switzerland AG 2019
R. Doss et al. (Eds.): FNSS 2019, CCIS 1113, pp. 172–189, 2019.
https://doi.org/10.1007/978-3-030-34353-8_13

1. Target and Decoy: These type of drones are mostly used for military purposes. It enables the controller who controls the drone to take a clear shot on a target by providing both ground and aerial views of surroundings.
2. Reconnaissance: Widely used for military purposes to collect battlefield intelligence and report to the operator.
3. Combat: also widely used in military battlefields to help carry missiles and to facilitate drone strikes.
4. Logistics: These types of drones help deliver cargos in hard to access areas, such as rural locations.
5. Research and Development: These are used for the purpose of developing drone technologies.
6. Civil and Commercial: These types of drones are mostly deployed during events to take photographs and to gather data.

Digital Forensics has become an important research topic in recent times, comprising the following sub-domains: network forensics, mobile forensics, database forensics, memory forensics, cloud forensics and data/disk forensics. Digital Forensics is considered to be the most complex stage of the cybercrime investigation process, to facilitate drawing of valid conclusions by investigators. It comprises the collection of data in a scientifically sound manner, analysis and preservation of data acquired in the form of digital media. According to [5], an investigator needs to consider the following steps as part of an in-depth forensic investigation:

1. Make sure that the original copy of the data is not altered while conducting investigation. Best practice is to make a duplicate copy of the evidences found and work on them, instead of working on original copy.
2. The data which is gathered should be clean from malware, to make sure it does not affect the whole process of investigation.
3. The investigator must present data and report facts without being biased.

Self-Organizing Maps (SOM) are an unsupervised learning neural network model [17] that facilitates the organisation of data into clusters, to help in decision making. We have adopted SOM for carrying out a digital forensic investigation on drone data, through application of correlation rules; identifying and bifurcating data into groups and ascertaining patterns to visually present the data clusters to the digital forensic investigator, so as to help predict patterns or behavior in the data clusters [17]. The maps that are generated in SOM exhibit numerous/advanced visualizations of large higher-dimensional datasets. With the help of these visualizations investigator can locate the data/information more easily and effectively from a large amount of data [18]. Based on the clustering results obtained we can create visualizations and plot the graphs and reverse engineer the data to help identify what type of data went into which cluster, so that the forensic investigator can easily traverse into that cluster to predict the behavior of the drone/flight, which was involved in a suspicious activity.

The rest of the paper is organized as follows: a background on digital forensics and drones forensics is presented in Sect. 2. Section 3 presents the proposed method for forensic investigations. In Sect. 4, we define the dataset used for the two drones that were tested for this research. In Sect. 5, we detail the steps taken through application of

SOM on the drone datasets. Section 6 presents the results obtained from experiments, and Sect. 7 concludes the paper.

2 Background

Drones, aka unmanned aerial vehicles, are increasingly being deployed to effect a large domain of applications including the delivery of packages to remote locations, taking pictures of critical events, to fire missiles in a battlefield, and in agriculture, for effective delivery of resources

2.1 Digital Forensics

The main motto of digital forensics is to help investigators conduct an in-depth investigation by following a clear and concise investigation process model, while documenting any digitally stored evidence, and to derive conclusions. The whole process of digital forensics can be broken down into 5 stages and each and every stage may further divided into multiple sub stages [17]:

1. Identifying the Data for evidence:
   This stage comprises steps for identification of any digital or physical evidence present at the crime location. It is very important to identify the scope of actions before initiating investigating the case. Questions to be asked include: who are responsible for the case? And what are the best sources for the potential evidence? how many and what type of devices were involved in the crime, etc. Considering drones as an investigation component, evidence can be found on controllers, camera, GPS module and some peripheral devices such as flash drives, scanners, cloud storage, memory cards, computers/mobile phones, or any other electronic device that stores data. We can acquire data that includes the location, payload weights, flight logs, GPS ports, deleted records, flight controller settings, and inputs.
2. Acquiring/Collecting the data identified:
   Acquiring or collecting data should be done in the most appropriate manner in such a way that the integrity of the data should not be lost during the forensic analysis. It is very important to make sure that the digital evidence should be the exact evidence of the original copy and further cryptographic hashing algorithms were implemented on the copy of the original file to make sure that the data is not altered [5]. Volatile and non-volatile are the two main sources of evidence and both possess different methods for acquisition. According to [5], after data acquisition from the crime scene, one must make sure that the data meets the requirements mentioned below:

- Admissible
- Authentic
- Complete
- Reliable and
- Believable

3. Preservation of the data acquired:

Data sources should be preserved in such a way that the evidence gathered should remain unaltered, even after the whole process of investigation has been completed. According to [6], this can be achieved with the help of well-defined processes aka chain of custody, which means one can clearly know when, where, who, why and how the evidence is used in the process of investigation, to make sure that the evidence remain safe from unintended modifications carried out by an adversarial party.

4. Examining/analyzing the preserved data:

This stage plays a key role in the process of investigation; comprising a rigorous use of forensic tools after conducting a careful study of multiple forensic models in order to make sure that the forensic process is on the right track. Categorizing the available tools for a given dataset, is also a critical requirement for success in examination of the data. Care must be taken when analyzing the data, as intricate conditions such as the difference between the instance data and metadata emerging from an image or an email, can take the evidence analysis process in a wrong investigation direction. According to [6], during the process of analysis, commonly adopted stages are:

1. Verification of the Occurrence
2. Description of the System
3. Analyzing Time-Frame
4. Categorizing the Data
5. Filtering the Data acquired using Keyword search
6. Recovering Deleted Files/Data
7. Capturing of Live RAM Dump
8. Creating Disk Image
9. Discovery and Final Analysis

5. Reporting:

This is considered to be the last stage of the investigation process. During this stage a proper documentation which depicts the information about the methods adopted, tools used, the process involved and the results acquired, should be clearly presented. For a sound forensic investigation, as a forensic investigator one should design/implement a clear and concise digital forensic process model which must include both the evidence deriving methodology and the associated technical approaches adopted. According to [7], it is stated that the Digital Forensic Process Model (DFPM) should possess the following criteria:

- Meaning: Ensures that the actual meaning and the interpretation of the digital evidences were not affected in the process of forensic investigation.
- Errors: Ensures that all the errors in the process were clearly identified and explained, so that it does not affect the process.
- Transparency: Ensures that all the work being done in the process of investigation is transparent and examined independently.
- Experience: Ensures that the forensic investigator who has undertaken the process has got handful knowledge on digital forensic investigation.

2.2   Digital Forensics for Drones

In [1], a method is proposed to access the digital storage medium on drone memory, comprising three main stages namely acquisition, analysis and reporting. This proposed method combines file/data carving and forensic analysis of operating system and image/EXIF optical sensor metadata, video files, data obtained from remote sources and flight path data based on GPS coordinates and controller chip information. The method was tested on the Parrot AR Drone 2.0. In [8], an enhanced digital investigation process model is presented. The model consists of five phases, namely, readiness, deployment, physical crime scene investigation, and digital crime scene investigation. The main motto of these phases is to collect the gathered evidence from the crime scenes either physically or electronically, and to start the analysis based on the acquired data. However, the proposed model does not provide a clear differentiation between the victim and the suspects of a primary crime investigation. Also, by looking at the 3rd and 4th phases of the integrated model, one cannot confirm whether it is a digital crime or physical crime investigation, until and unless some preliminary investigation is executed. To eliminate the mentioned flaws in the above method, in [8], the authors present an enhanced integrated digital investigation process model to expand the deployment phase, comprising a merge between both physical and digital crime investigation phases, and by introducing a trace back phase, where criminal physical crime scene is traced down to help trace down devices that were used in the crime scheme.

In [9], the authors present a methodology that is well suited for all genres of digital forensics. An abstract model for forensic investigation is developed by comparing four different forensic methodologies, which exist in the literature. Their proposed model consists of 9 stages, namely, identification, preparation, approach strategy, preservation, collection, examination, analysis, presentation and returning evidences. This proposed model does not focus on tools and technologies used for the investigation, but rather addresses the standardized methodology for any digital forensic genre. It is also stated that this model is best suited in order to deal with past, present and future digital technologies in a well understood and disciplined manner, which is considered to be one of the finest advantages. But this could also mark as a disadvantage because the model is too generic and was not tested on any specific application or technology domain.

In [10], the authors have addressed the challenge of reporting of forensic findings. It is stated that the reports generated from some of the tools are not in forensically sound manner and are not admissible in the court of law. The authors adopted a 20 step approach to identify the make and model of the drones. All the steps include identifying information about the hardware or software used, about the data storage locations, acquiring digital or physical evidences and so on. The Encase tool is used to perform file carving and to analyze dat files. This gives information only about the dates when the file was created, so in order to interpret the data they have used Datcon, which yields data on GPS coordinates, battery capacity, height and KML formats of the flight path.

In [11], a framework is presented for examination of flight logs of micro drones. They have chosen to work on three micro drones, namely, 3DR Solo, Yuneec

Typhoon H and DJI Phantom 4. They have mentioned that the logging system on the drones completely depends on the autopilot software and the board. Every different drone manufacturer stores log data in different formats. The authors studied the details of log files in various types of drones. They have also addressed how important it is to choose the proper drone because some of the drones are not capable of recording flight logs. Their proposed model for digital forensics is not best suited for all the drones because as mentioned above, some of the drones are not capable of storing logs, hindering the digital forensic analysis of the data.

In [12], the authors presented a 7-phase framework for digital forensics, comprising, preparation, scene control, customization detection, data acquisition, evidence authentication, evidence examination and presentation. After acquiring the data they have used FTK Imager tool in order to authenticate the data they have attained by comparing their MD5 hashes of the files. They were able to acquire 3 forms of evidence such as.txt file which is generated by DJIGo application, contains information about the controller,.dat files from the drone SD card about the flight path information and EXIF files in the drone internal memory, comprising information on the images taken by the drone. In order to analyse the.txt file, they have used the online tool called airdata.com, which helps in converting.txt files into user understandable.csv formats. The proposed framework mainly regenerates the flight path of any drone that is meant for investigation but lacks in detecting classification and category of the suspicious drones. Improving upon this scheme, in [13], the ExifTool is deployed for analysis of EXIF files, and to analyse.dat files of the drones.

In [14], the researchers present a method that focuses only on acquiring GPS log data for presentation as digital evidence. In order to obtain GPS log data, the authors have chosen to investigate the DJI Phantom 3 Advanced drone, because GPS coordinates of DJI Phantom 3 are always stored as data within the flight logs. From their findings, it can be concluded that the percentage of digital evidence found in the storage of UAV is 50% and the evidence as found in the memory card is 16.6%.

3 Proposed Method/Technique

In contemporary times, with the help of drones many criminal offenses are being conspired, and it is very important to conduct a forensic analysis on drones that are involved in such incidents, in order to mitigate future threats. Researchers and forensic investigators alike have proposed generic methods to conduct a proper investigation on drones that are compromised. After conducting an in-depth research on digital forensics and the appertaining process of investigation, considering the methodology proposed by various researchers (see previous section), it can be noted that researchers basically were unable to address and accurately analyse the datasets that are obtained from the flight paths. Such analysis can portray a clear view on the behavior of drones and the actual nature of the threat. Lack of research in this particular domain i.e., forensic investigation on drone flight paths, has motivated us to carry out research on digital forensics of drone data through utilization of several well-known offline and online digital forensic tools. Choosing a right forensic tool is crucial, because a single drone

can hold many numbers of file systems in its memory, requiring the adoption of multiple tools to aid in the analysis.

Considering the above mentioned views on digital forensics, the holistic investigation process as envisaged by us would comprise three stages of investigation. The stages in forensic investigation of the proposed methodology are as follows:

1. Acquiring Data
2. Drone Forensic Investigation
3. Acquiring Results and Reporting (Fig. 1).

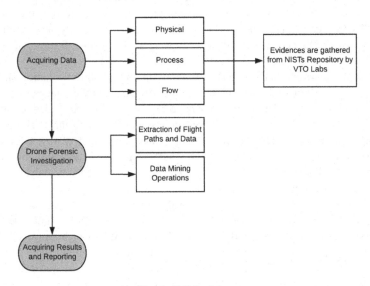

Fig. 1. Methodology

Acquiring Data:

The main purpose of this stage is to focus on where the data is and what are its sources? Such information will help us attain evidences that are existing in the device. In general, acquiring data can be done in 3 different ways:

1. Physical: gathering the evidence from physical devices like Drones, Sensors, Battery, Remote Control, etc.
2. Process: comprises information about how the operator runs the flight.
3. Flow: gathering evidence about the flow of communication between the drone and its vicinity/environment e.g. Wi-Fi Traffic, Cloud data etc.

However this research mainly focus on conducting a clear and concise forensic investigation on drone images which are extracted from the CFReDS project [19], all the evidence is stored in the form of forensic images in the repository, powered by VTO labs. So during the first stage we collect the evidence from the CFReDS repository and will start analyzing with the help of some forensic tools.

Drone Forensic Investigation:
Comprises analysis of the evidence which is gathered from the first stage using some of the available forensic tools, in order to know what type of data/evidences are available. This stage addresses the conversion of the data gathered into human readable formats, to gain information about the flight logs, paths etc. During this stage, some of well-known tools to read and convert the files into readable format are deployed. These include UTrack to attain flight paths, GPX Visualizer and GPXSee in order to read KMZ/KML and GPX files, Datcon, CSVView to read.dat files and convert them into CSV or txt format.

With the available evidence and files extracted and analyzed, as a further step, we adopt the SOM clustering technique to examine and cluster the drone data. The main focus during this stage is to retrieve the results based on the analysis on datasets with the help of clustering algorithms.

Acquire Results and Reporting:
This stage comprises steps to help draw conclusions based on the analysis of the acquired data, and to generate reports which include GNU Plots and Visualizations for future analysis. During this research in order to undertake the forensic investigation on drones, the above mentioned methodology was tested on the ArduPilot DIY and DJI Phantom 4 drones.

4 Dataset Description

Once the evidence were extracted from NISTs repository, based on the type of evidence, tools were chosen to extract valid flight paths and their associated datasets. This section showcase how the datasets of each valid flight path were obtained and what information it carries and sizes of each and every dataset.

Two different tools were tested in order to obtain datasets from the flight paths of ArduPilot DIY and DJI Phantom 4 drones. Because evidence found in the ArduPilot DIY is in the form of GPX files, the GPX Visualizer is deployed to extract flight paths and their datasets, whereas DJI Phantom 4 carries evidences in the form of.dat, therefore, the Air Data.com online tool is used to extract flight paths and datasets.

ArduPilot DIY Drone:
Out of 15 evidences found from the CFReDS repository, the first 11 files among 15 GPX files tested with GPS Visualizer and various tools and yielded zero data in relation to the flight patterns, so we identified the last 4 GPX files for our analysis, in order to attain the information about flight paths. Once the flight paths from each of these four files were obtained with the help of GPS Visualizer, datasets associated with each of the flight paths were also be extracted with the same tool; comprising the following information on flight paths:

- Type
- Time
- Latitude
- Longitude
- Altitude
- Speed measured in "Km/h"
- Distance measure in "Km"
- Distance Interval in "m"
- Pitch
- Roll

Each attribute represents how the flight as a whole is operated and the information gathered here is further utilized to assess the behavior of the flight, and to help forensic investigator apply efficient techniques such as clustering.

The size of each dataset is sizably small with 45, 75, 26 and 107 instances in each of the datasets 1, 2, 3 and 4 respectively. This indicates that the flight is operated over a short duration, and is holding very less information. The size of the dataset plays an important role during further analysis through clustering. The more the data, higher is the intuition of each cluster and behavior of the flight.

DJI Phantom 4:
Unlike the ArduPilot DIY drone, evidence drawn from this drone are in the.dat file format. Among 37 dat files, 14 files did not yield valid evidence, and tests were conducted on rest of the 23 files, which yielded sufficient information in relation to the flight path. Flight Numbers that were not carrying any valid GPS data are as follows: 14, 17, 18, 28–37, 40. An online tool, namely, AirData.com was used in order to attain the flight paths and the datasets associated to them.

The datasets retrieved from the flight paths of DJI Phantom 4 are large in size and carry more useful information to support the investigation. Information comprised power consumption, sensors and when they were used, controls, weather and media are also associated with the file.

Some of the useful attributes that each dataset of DJI Phantom 4 holds is as follows:

- Date and Time
- Latitude
- Longitude
- Voltage (v)
- Max Altitude (feet)
- Max Ascent (feet)
- Max Speed (mph)
- Max Distance (feet)
- Pitch (degrees)
- Roll (degrees)

- Current (A)
- Battery temperature (f)
- Fly State Raw

Unlike ArduPilot DIY drone, datasets obtained from the flight paths of DJI Phantom 4 depict that the flight is operated in multiple modes, data from most of the flight paths are recorded when the flight is operated in Motors Started, Assisted Takeoff and P-GPS modes. Other flight modes are Assisted Takeoff, Atti, Go Home, Auto Landing and Confirm Landing.

Dataset for file FLY003 recorded data when the flight is in Motors Started, Auto Takeoff and P-GPS mode, whereas, datasets for files FLY 004, 005, 007, 008, 009, 011, 013, 015, 020, 023, 024, 025, 027 hold the data when the flight is in Motors Started, Assisted Takeoff and P-GPS mode. On the other hand FLY 006 and FLY 026 recorded data when the flight is in Motors Started, Assisted Takeoff, P-GPS and Auto Landing state. FLY 010 when the flight is in Motors Started, Assisted Takeoff, P-GPS and Go Home states. FLY 012 holds the data when the flight is in Motors Started, Assisted Takeoff, P-GPS, Go Home and Auto Landing position. Dataset of FLY 016 holds the data when the flight is in only P-GPS mode. FLY 019 holds the data when the flight is in Motors Started, Assisted Takeoff, P-GPS, Go Home, Auto Landing and Confirm Landing mode. Dataset of FLY 022 holds information when the flight is in Motors Started, Assisted Takeoff, P-GPS and Atti mode. And finally datasets of FLY 038 and 039 hold the data when the flight is in Motors Started, Auto Takeoff, P-GPS, Auto Landing and Confirm Landing, when the flight is in Motors Started, Auto Takeoff, P-GPS and Auto Landing respectively.

Because the size of the datasets are large, this reduces the complexity while deriving meaning from the clusters obtained through the application of SOM on each of the above files.

5   Forensic Analysis Steps Taken/Procedure

For quick understanding on how the actual process of digital forensic investigation on drones has done, the entire process is represented in the form of flowcharts and are depicted below with a detailed overview.

The digital forensic investigation process is further segregated into two stages, as follows:

1. The first flowchart represents the actual process of gathering evidence, extraction of data and flight paths (Stage 1).
2. The second flowchart represents the final stage of analysis, which is known as clustering, and depicts how the data is clustered with the help of SOM (Stage 2).

Stage 1:

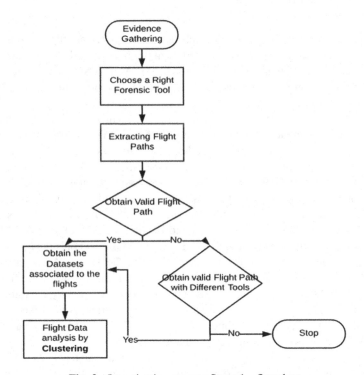

Fig. 2. Investigation process Stage 1 - flowchart

As the first step of investigation, it is always important to collect the evidence as found from the source. In this investigation, for both the drones ArduPilot DIY and DJI Phantom 4, evidences were gathered from log files and forensic images of the NIST repository. Identifying the right forensic tool plays quite an important role in the investigation process, and helps guide the investigation through information on what type of data/evidence is available in order to move forward. Different tools were chosen based on their accuracy in carrying out flight path extractions. GPS Visualizer and UTrack are the two main forensic tools deployed to extract flight paths and data associated with each and every path, of the ArduPilot Drone, whereas AirData.com is the main tool chosen to undertake the forensic investigation on the DJI Phantom 4 drone (Fig. 2).

During investigation, based on evidence/files acquired, some of them were unable to draw any valid data, so we tried extracting valid flight paths from the data using several other tools. GPXSee, Track Detective, CsvView, and DatCon are the alternate tools that were tested to acquire data from ArduPilot DIY and DJI Phantom 4 drones, respectively. Once the valid flight paths from both the drones were extracted, we obtain the datasets associated with individual flight paths. These datasets carry information

about flight logs: latitude, longitude, altitude, speed, distance, time, date, pitch, roll etc. From these datasets, further analysis is conducted based on clustering the data (SOM).

Stage 2:
See Fig. 3.

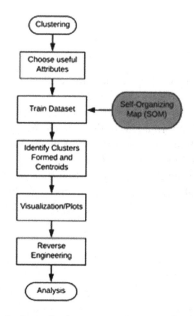

Fig. 3. Investigation process Stage 2 - flowchart

Clustering:

Clustering is nothing but a technique or process of grouping data points into small individual groups in such a way that data points in the group are more similar to peer data points of the group, when compared to the data points of other groups. The main motto behind clustering is to find out the data patterns in high dimensional unlabeled data. Consequently, clustering helps apply unsupervised pattern matching to group data based on similarity. As a first step in order to cluster the extracted datasets obtained from the forensic tools, finding out the right data mining tool is vital for effective clustering. We therefore chose the popular data mining tool, namely, Weka, for SOM application to the droned datasets. SOM is known to cluster data based on a high degree in accuracy for similarity matching between elements of a single cluster.

Weka is a well-known data mining tool that provides comprehensive data mining algorithms and data preprocessing techniques that enable the application of data mining algorithms to datasets. Weka algorithms for regression, classification, attribute selection, clustering and association rule mining enable the user to quickly explore the dataset [15]. With its well-designed graphical user interface, it supports data mining tasks such as [16]:

1. Data Preprocessing
2. Classification
3. Clustering
4. Regression
5. Feature selection and
6. Visualization.

6   Experimental Results

Preliminary tests were conducted to validate the credence of files obtained from the drones. Out of a total of 15 GPX files, 4 were found to contain valid flight paths for the ArduPilot DIY drone. For the DJI Phantom 4 drone, 23 out of 37 dat files were found to contain valid flight path information. In this section, we present the results obtained from the analysis of the flight path information of the two drones.

Plots that are illustrated in Figs. 4, 5, 6 and 7, are generated based on the results obtained when the 4 valid flight datasets were trained with the SOM clustering algorithm, for the ArduPilot DIY drone. The x-axis represents the instance number of the GPS coordinate in the flight path, and the y-axis represents the corresponding latitudes.

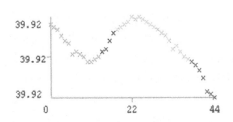

Fig. 4. ArduPilot DIY drone dataset 1 plot

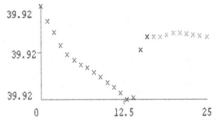

Fig. 5. ArduPilot DIY drone dataset 3 plot

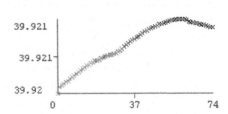

Fig. 6. ArduPilot DIY drone dataset 2 plot

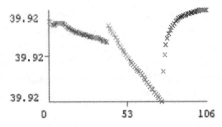

Fig. 7. ArduPilot DIY drone dataset 4 plot

Plots (Figs. 8, 9, 10, 11 and 12) will represent the results that are obtained when the valid flight datasets were further trained in Weka by applying SOM as a clustering algorithm. The 23 valid flight paths FLY (003–013), (015–016), (019–027), (038–039) clustering results in the form of graphs or plots were depicted clearly and their axis labels are as follows (Figs. 13, 14, 15, 16, 17, 18, 19, 20, 21, 22, 23, 24, 25, 26, 27, 28 and 29).

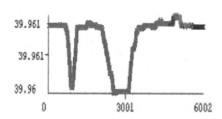

Fig. 8. DJI Phantom 4 FLY003 plot

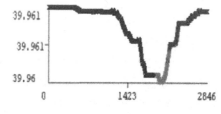

Fig. 9. DJI Phantom 4 FLY006 plot

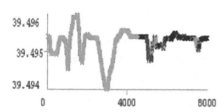

Fig. 10. DJI Phantom 4 FLY004 plot

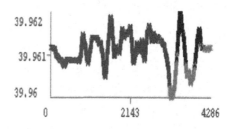

Fig. 11. DJI Phantom 4 FLY007 plot

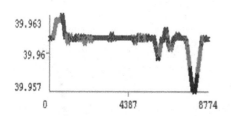

Fig. 12. DJI Phantom 4 FLY005 plot

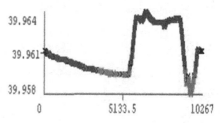

Fig. 13. DJI Phantom 4 FLY008 plot

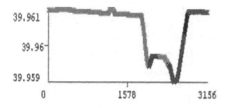

Fig. 14. DJI Phantom 4 FLY009 plot

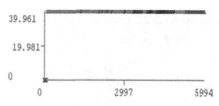

Fig. 15. DJI Phantom 4 FLY010 plot

Fig. 16. DJI Phantom 4 FLY011 plot

Fig. 17. DJI Phantom 4 FLY012 plot

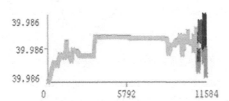

Fig. 18. DJI Phantom 4 FLY015 plot

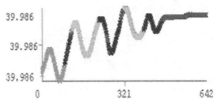

Fig. 19. DJI Phantom 4 FLY016 plot

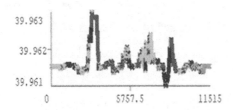

Fig. 20. DJI Phantom 4 FLY019 plot

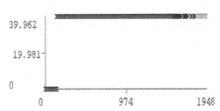

Fig. 21. DJI Phantom 4 FLY020 plot

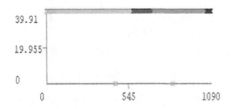

Fig. 22. DJI Phantom 4 FLY022 plot

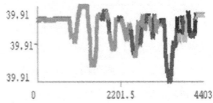

Fig. 23. DJI Phantom 4 FLY023 plot

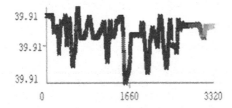

Fig. 24. DJI Phantom 4 FLY024 plot

Fig. 25. DJI Phantom 4 FLY025 plot

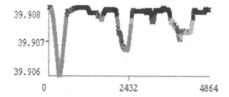

Fig. 26. DJI Phantom 4 FLY026 plot

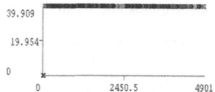

Fig. 27. DJI Phantom 4 FLY027 plot

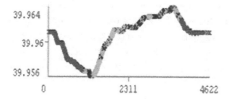

Fig. 28. DJI Phantom 4 FLY038 plot

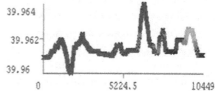

Fig. 29. DJI Phanton 4 FLY039 plot

Analysis:

Based on the analysis of the above plots, we observe a major operational difference in both the drone flight paths, and a large amount of artefacts were also found associated with DJI Phantom 4 when compared to ArduPilot DIY drone. This was as predicted, as the DJI Phantom 4 is considered to be more sophisticated device with more sensors and a high resolution camera, with the data being recorded when the flight is operated in many different modes, namely, Motors Started, Assisted_Takeoff, Auto Takeoff, P-GPS, Atti, Go Home, Auto Landing and Confirm Landing.

Based on the results obtained for both the drones by applying the SOM clustering algorithm on the respective datasets, results obtained from the DJI Phantom 4 are more informative and enable the investigator to predict the phases of the flight journey based on the clusters of individual data points, whereas the results obtained for the datasets of ArduPilot DIY drone are not very informative and elude towards vague patterns in flight path data, which results in more complexity when deriving intuition from the cluster information.

Moreover, each dataset in ArduPilot carries less data when compared to the data obtained for the DJI Phantom 4, which could also be another reason that data in DJI Phantom 4 is more informative.

In DJI Phantom 4, the clusters in most of the data sets were formed in a way that, the one which holds majority of the instances dictates the initial phases of the flight path and the one which holds the highest distance, speed, altitude and ascent, including high battery temperature depicts the final phases of the flight journey. And similarity between the clusters is also fairly high as the flightpath attributes including distance, speed, altitude and ascent, are similar across the various datasets of the drone, whereas for the ArduPilot DIY drone, the dataset is very small (4 files all together), and the similarity between the clusters is significantly high, consequently encumbering the pattern identification task.

7 Conclusions and Future Work

It is a known fact that today many UAVs or Drones are being hijacked by intruders in order to conduct malicious activities using drones. Digital forensic investigation on affected drones is therefore imperative. This paper addresses the problem through application of a known clustering technique, namely, the self organizing map, in order to support digital forensic investigation of drone data. The proposed methodology is tested on two drones, namely, ArduPilot DIY and DJI Phantom 4. As a part of investigation, flight paths were extracted and their associated datasets were obtained from both the drones and these datasets were further subjected to SOM-based clustering using the Weka tool. Results obtained identify DJI Phantom 4 drones to hold more evidence and they are forensically sound when compared with ArduPilot DIY drones. As part of our future work, we intend to assess the tools required for forensic investigation of drone flight paths, as well as on extraction on other evidence types as found in drones including chip-off data.

References

1. Bouafif, H., Kamoun, F., Iqbal, F., Marrington, A.: Drone forensics: challenges and new insights. In: 2018 9th IFIP International Conference on New Technologies, Mobility and Security (2018). https://doi.org/10.1109/ntms.2018.8328747
2. Mazurczyk, W., Caviglione, L., Wendzel, S.: Recent advancements in digital forensics, part 2. IEEE Secur. Privacy **17**(1), 7–8 (2019). https://doi.org/10.1109/msec.2019.2896857
3. Howell, E.: What Is A Drone? Space.com (2018). https://www.space.com/29544-what-is-a-drone.html. Accessed 15 Mar 2019
4. What are Drones? Hobbytron.com (2019). https://www.hobbytron.com/lc/what-are-drones.html. Accessed 15 Mar 2019
5. Reyes, A., O'Shea, K., Steele, J., et al.: Digital Forensics and Analyzing Data. Cyber Crime Investigations, pp. 219–259 (2007). https://doi.org/10.1016/b978-159749133-4/50010-3
6. Rana, N., Sansanwal, G., Khatter, K., Singh, S.: Taxonomy of Digital Forensics: Investigation Tools and Challenges. eprint arXiv:1709.06529 (2017)
7. Chernyshev, M., Zeadally, S., Baig, Z., Woodward, A.: Mobile forensics - advances, challenges and research opportunities. IEEE Secur. Privacy **15**(6), 42–51 (2017). https://doi.org/10.1109/msp.2017.4251107
8. Baryamureeba, V., Tushabe, F.: The enhanced digital investigation process model. In: Digital Forensic Research Conference (2004)
9. Reith, M., Carr, C., Gunsch, G.: An examination of digital forensic models. Int. J. Digit. Evid. **1**(3), 1–12 (2002)
10. Rodor, A., Choo, K., Le-Khac, N.: Unmanned Aerial Vehicle Forensic Investigation Process: DJI Phantom 3 Drone As A Case Study. arXiv.org. arXiv:1804.08649 (2018)
11. Renduchintala, A., Albehadili, A., Javaid, A.: Drone forensics: digital flight log examination framework for micro drones. In: International Conference on Computational Science and Computational Intelligence (CSCI) (2017). https://doi.org/10.1109/csci.2017.15
12. Gulatas, I., Baktir, S.: Unmanned aerial vehicle digital forensic investigation framework. J. Naval Sci. Eng. **14**(1), 32–53 (2018)

13. Barton, T., Azhar, M.: Forensic analysis of popular UAV systems. In: 2017 International Conference on Emerging Security Technologies (EST) (2017). https://doi.org/10.1109/est. 2017.8090405

14. Prastya, S., Riadi, I., Luthfi, A.: Forensic analysis of unmanned aerial vehicle to obtain GPS log data as digital evidence. Int. J. Comput. Sci. Inf. Secur. **15**(3), 280–285 (2017)

15. Hall, M., Frank, E., Holmes, G., et al.: The weka data mining software: an update. ACM SIGKDD Explor. Newslett. **11**(1), 10–18 (2009). https://doi.org/10.1145/1656274. 1656278

16. Singal, S., Jena, M.: A study on WEKA tool for data preprocessing, classification and clustering. Int. J. Innov. Technol. Explor. Eng. **2**(6), 250–253 (2013)

17. Fei, B., Eloff, J., Venter, H., Olivier, M.: Exploring forensic data with self-organizing maps. In: Pollitt, M., Shenoi, S. (eds.) DigitalForensics 2005. ITIFIP, vol. 194, pp. 113–123. Springer, Boston, MA (2006). https://doi.org/10.1007/0-387-31163-7_10

18. Feyereisl, J., Aickelin, U.: Self-Organizing Maps in Computer Security. arXiv.org, arXiv: 1608.01668 (2016)

19. https://www.cfreds.nist.gov/. Accessed 15 July 2019

An Investigation of Using Time and Ambient Conditions to Sense the Unexpected Removal of RFID Tag

Yuju Tu[✉]

National Chengchi University, Taipei, Taiwan (ROC)
tuyuju@nccu.edu.tw

Abstract. The unexpected RFID tag removal is a severe problem to RFID-based information system. It can result in erroneous asset identification, mismatched item-level information, and such security threats as impersonation or identity theft. However, only very few scholars and practitioners have ever carefully investigated this problem and managed to solve it. On the other hand, many firms are going to tag their overall assets, products, etc. with RFID tags, but they are currently either clueless about the problem or limited in the choices for solving it. This study presents a solution prototype of integrating time and ambient conditions to sense the RFID tag unexpected removal in an autonomous and secured way.

Keywords: RFID · Ambient conditions · Tag removal · Time

1 Introduction

It is well established that use of RFID (Radio-Frequency Identification) is a very efficient and cost effective way to manage assets. Nowadays, there is a variety of information systems based on RFID for supporting asset management in firms, supply chains, etc. Moreover, RFID-based information system is generally considered a very promising IoT (Internet of Things) solution. This is largely because RFID tag has many merits including the extra-ordinary benefit of being employed and deployed without concerning the issues of battery, electricity cable, and so forth. One important premise of using RFID-based information system to achieve asset management is that the system can correctly provide asset identity information, such as location. However, once RFID tag is unexpectedly removed from its associated asset, the RFID-based system will soon fail in asset identification.

For preventing the failure, the main research objective of this study is to investigate the solution in an autonomous and secured manner for dealing with the tag unexpected removal problem. In many regards, this problem is a severe threat in today's world [1, 2]. For example, when a contactless transportation pass card's RFID tag is unexpectedly removed, this card generally has no use, because the tag is equivalent to the card holder's identity. Relatedly, when an apparel's RFID tag is unexpectedly removed, the apparel is essentially out of the control of the associated RFID-based system. More seriously, the unexpectedly removed RFID tag is possible to be misused for impersonating another

R. Doss et al. (Eds.): FNSS 2019, CCIS 1113, pp. 190–197, 2019.
https://doi.org/10.1007/978-3-030-34353-8_14

asset's or person's identities. For example, it is not uncommon that the price tag of a cheaper asset is maliciously removed and then reattached to another luxury asset [3]. In other words, the RFID tag unexpected removal is a very challenging and important topic to address.

This study focuses on proposing the informational solution, rather than the mechanical or manual solution, for addressing the unexpected RFID tag removal problem. In practice, several seemingly-related manual solutions have been proposed and used in many places, such as electronic article surveillance (EAS) system or Ink-tag system. US NASA recently proposed a mechanical solution based on a set of customized devices for checking if RFID tag is well placed in cargo [3, 4]. Nevertheless, these solution are specifically designed for preventing asset shoplifting or damage, rather than for immediately sensing the unexpected removal of RFID tag.

In short, the overall contribution of this study is multi-fold. First, it enriches and broadens the research area for handling the RFID tag unexpected removal problem. In prior literature, extremely few studies have ever investigated this problem [28]. However, this problem is an obvious concern to thousands of firms that are overall investing a huge amount of dollars in RFID to collect the item-level information [1, 2, 4, 5]. Moreover, this study presents the prototype of a lightweight and easy-to-use solution. It is compatible with ISO EPC C1 G2 standards and applicable to even the most resource-constrained RFID-based information system. More importantly, this study can largely increase the general cognizance level to deal with the problem of RFID tag unexpected removal.

In the following sections, this paper presents the major idea of this study to sense the RFID tag unexpected removal by integrating time and ambient conditions. Initially, this paper details the problem of losing identification of the RFID tagged asset. Next, this paper provides a concise review of the related literature. This paper then illustrates a prototype for instantiating the proposed idea of this study for handling the RFID tag unexpected removal problem. This paper also provides an analysis of the security characteristics of the prototype. Finally, this paper ends with a brief discussion of the main value, limitation, and conclusion of this study.

2 The Problem of Losing Asset Identification Due to Tag Unexpected Removal

The loss of asset identification is a severe risk in asset management, because the loss of identity information often denotes the loss of the control of asset. Nowadays, many firms use RFID tags as their assets' identifiers for managing these assets. Each of these tags contain the unique identification for each asset in these firms. Thus, the unexpected tag removal is one main cause of losing asset identification. For example, the loss of a product's identifier, such as a price tag, often means not being able to sell the asset to customer. Generally, the loss of asset identification is mainly due to two reasons. The first one is accidental and the second one is intentional. For instance, it is not

uncommon that a package's RFID tag is unexpectedly removed because the package is accidentally dropped from a higher place to the ground or because the tag is intentionally removed by a person.

Moreover, any present RFID-based information system is very vulnerable to the unexpected tag removal. There are considerable examples for showing that [6–9]. One example is that a person is caught in a US shop for removing a price tag from one cheap asset and then putting it on another expensive asset in order to pay less for more. Similarly, several people are caught in Wal-Mart or other well-known shops for switching the price tags on many valuable assets, such as electronic consumer products, and bringing back the assets without the tags to the shops for refund. On the other hand, RFID tag is very difficult to be manufactured, as compared with the common bar code. Thus, it is conventionally said that people barely can fool the RFID-based system for purchasing the tagged asset at a lower cost. However, removing an RFID tag from any asset and then sticking it with one from a more valuable asset is relatively easier when the RFID tag is embedded in a removable object like a hanging price tag. This means that the RFID-based system is very vulnerable against the unexpected tag removal in the reality. In the same vein, the unexpected tag removal is likely to result in the very bad supply chain performance, because of inventory shrinkage, inconsistency, discrepancy, and so forth between upstream and downstream supply parties. In prior literature, many studies have the investigations for solving such supply chain management problems [1, 2, 11, 12], but almost all of them ignore a fact that the unexpected RFID tag removal occurs. In practice, if the removal problem is not well addressed, the overall item-level information based on the RFID tagged assets will never be reliable and thus in turn result in very bad performance with respect to asset management, supply chain management, etc.

3 A Review of Related Literature for Handling the Unexpected Tag Removal

In prior literature, two steams of studies are relevant to what this study is fundamentally after. The first stream is based on using time and the other stream is focused on using ambient conditions [28]. For example, several RFID distance-bounding protocols use time to achieve proximity-checking [14–16, 18, 19]. This is because time always clicks and changes in any moment. In other words, time is a suitable indicator for marking change. For example, if any RFID signal is maliciously passed on, it is likely to take a longer time to receive the signal. In other words, use of such a change of time is suitable for detecting RFID relay attack. However, what RFID distance-bounding protocols are mainly focused on is the round trip time (RTT) of RFID signal, but RTT is not directly associated with the RFID tag removal. Moreover, RTT is very sensitive to the measurement unit of time, because RFID signal transmission speed is similar to the speed of light transmission. In other words, use of RTT would not be suitable for detecting the RFID tag unexpected removal. However, it is for sure that the removal essentially is a process and must need a series of time moments to take place.

Recently, several studies use ambient conditions to enhance RFID location awareness capability. For example, Ma et al. [21] use GPS to determine the distance between RFID tag and reader. Halevi et al. [17] use human factors, such as sitting posture and walking posture, to achieve proximity-checking. Halevi et al. [23], Urien and Piramuthu [15], and Tu and Piramuthu [33] use natural factors, such as sound, light, temperature, and magnetic field to achieve proximity-checking. Gurulian et al. [31, 32] use artificial factors, such as infrared light and vibration, to examine the distance between RFID tag and reader.

In particular, Tu et al. [28] propose a knowledge-based solution for detecting the separation of RFID tag from its associated object. Their solution comprises associated tags, knowledge-based system, and ambient conditions. On the other hand, their solution has not investigated the factor of time in the unexpected RFID tag removal.

4   The Proposed Solution of Integrating Time and Ambient Conditions

After reviewing the related solutions in prior literature, this study proposes using both time and ambient conditions to detect the unexpected tag removal for improving asset identification and security (Fig. 1).

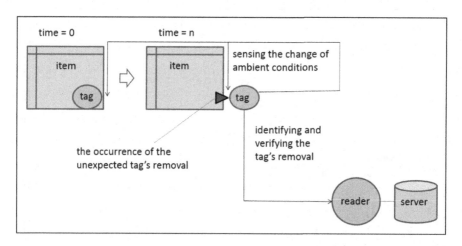

Fig. 1. The main idea of the proposed solution

Moreover, the following notations (Fig. 2) and diagram (Fig. 3) show the proof of concept (POC) of the proposed solution.

- $A_{T(t=1..n)}$: Tag's sensed ambient condition at the time (t) moment of 1, 2, to n

- $A_{R(t=1..n)}$: Reader's sensed ambient condition at the time (t) moment of 1, 2, to n

- $Corr()$: Correlation function

- $H_w()$: Hamming weight function

- $M_{T(i)}$: The tag's i_{th} generated temporary vector

- $N_{T(i)}$: The tag's i_{th} generated random nonce

- x: A shared secret for reader and tag

- y: A value variable

- ϵ: The maximum allowed value for correlation deviation

- $\oplus$: Exclusive-OR (XOR) operator

- $||$: Concatenation operator

Fig. 2. Notations

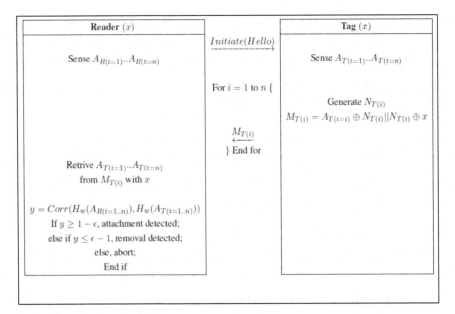

Fig. 3. A protocol for instantiating the proposed solution

Briefly stated, this prototype is based on a premise that a pair of reader and tag share with each other a secret (x) and both of them can sense their ambient conditions at a certain period of n time moments. In the beginning, the reader initiates the process of

checking the tag situation (e.g., removal or not) by sending a message of "Hello" to the tag. Afterwards, the reader and tag start to collect the readings of their ambient conditions. Next, the tag goes through a loop for n rounds. In the each round, the tag generates a fresh nonce and then packages it with the shared secret and each of its sensed ambient conditions of n time moments. The tag then subsequently sends to the reader each of the packages. After receiving the packages, the reader uses the shared secret to retrieve the tag-sensed ambient conditions of n time moments. Finally, the reader compares the ambient conditions of n time moments that are sensed by itself and by the tag, using a correlation function with the input of the Hamming weights of the two conditions. Moreover, if the result (y) shows that the reader-sensed and tag-sensed ambient conditions are significantly and positively correlated, it suggests that the tag likely remains attached. If the result shows that they are not correlated, it suggests that the tag is likely to be removed unexpectedly. If the result shows that they are not significantly correlated, it suggests that the tag is likely not responding normally for such reasons as false read, security attack, and so forth; the reader then thus aborts the process.

Moreover, this prototype has the following security characteristics on the basis of general RFID security evaluation criteria by Avoine et al. [30].

- Integrity and secrecy:
  This protocol will not reveal any private information in clear text because of the shared secret. In this way, it can keep the secrecy of the message. This means that authenticity of the communicating parties can be sustained constantly.
- Anonymity:
  Following the characteristic above, it is difficult to trace and track either the tag or reader because of the encrypted secret and nonce.
- Forward and backward security:
  Following the characteristic above, even if the shared secret is exposed to adversary for any unpredictable or unknown reasons, adversary still needs accurate and timely ambient conditions and nonce to break the prototype.
- Synchronization or desynchronization security:
  Following the characteristic above, this is not a concern because the prototype does not update anything on either reader or tag side in each round (process).
- Replay security:
  Following the characteristic above, because nonce and ambient conditions change in every round (process), replay attack is preventable to a certain degree.
- Relay security:
  Following the characteristic above, this prototype can use the correlation of ambient conditions to achieve proximity-checking. Thus, Mafia relay attack and terrorist relay attack are preventable to a certain degree.

5 Concluding Remark

In conclusion, this study is aimed at raising the general awareness of the unexpected removal of RFID tag problem. Although many news and anecdotes have clearly shown that the RFID tagged assets are very easy to lose their identifications, only very few

scholars and practitioners have ever carefully investigated this problem and managed to solve it. In the foreseen future, thousands of firms will tag their every assets, products, etc. with RFID tags, but they are currently almost clueless about the problem and limited in the choices of preventing it. Strictly speaking, this paper only presents the preliminary findings. For example, the suitable ambient conditions for sensing the RFID tag unexpected removal remain not very clear in this paper. However, because of this paper, more attentions will be paid to designing and developing superior solutions for handing the problem, in turn improving the overall performance of the RFID-enabled asset management, supply chain management, and so forth.

References

1. Zhou, W., Piramuthu, S.: Preventing ticket-switching of RFID-tagged items in apparel retail stores. Decis. Support Syst. **55**, 802–810 (2013)
2. Zhou, W., Piramuthu, S.: Identification shrinkage in inventory management: an RFID-based solution. Ann. Oper. Res. **258**, 258–300 (2015)
3. http://www.foxnews.com/us/2017/11/09/florida-woman-buys-walmart-electronics-worth-1800-for-3-70-police-say.html. Accessed December 2017
4. https://technology.nasa.gov/patent/MSC-TOPS-49
5. Frost, S.: RFID market in apparel supply chain, May 2012
6. DTechEx, Apparel RFID 2013–2023, June 2017
7. http://www.schneier.com/blog/archives/2008/10/upc-switching-s.html (2008)
8. http://www.schneier.com/blog/archives/2012/05/bar-code-switch.html (2012)
9. McMillan, R.: Tech. Exec. Built Stolen 'Legoland' in $2M Home. Wired, 25 May 2012
10. Schuman, E.: Wal-Mart Stung in $1.5 Million Bar-Code Scam. eWeek, January 5 2005
11. Zhou, W.: RFID and item-level information visibility. Eur. J. Oper. Res. **198**(1), 252–258 (2009)
12. Atali, A., Lee, H., Ozer, O.: If the inventory manager knew: value of visibility and RFID under imperfect inventory information. SSRN 1351606 (2009)
13. Brands, S., Chaum, D.: Distance-bounding protocols. In: Helleseth, T. (ed.) EUROCRYPT 1993. LNCS, vol. 765, pp. 344–359. Springer, Heidelberg (1993). https://doi.org/10.1007/3-540-48285-7_30
14. Hancke, G.P., Kuhn, M.G.: An RFID distance bounding protocol. In: Proceedings of the IEEE/Create-Net SecureComm, pp. 67–73 (2005)
15. Urien, P., Piramuthu, S.: Elliptic curve-based RFID/NFC authentication with temperature sensor input for relay attacks. Decis. Support Syst. **59**, 28–36 (2014)
16. Tu, Y., Piramuthu, S.: RFID distance bounding protocols. In: 1st International EURASIP Workshop on RFID Technology, pp. 67–68 (2007)
17. Halevi, T., Lin, S., Ma, D., Prasad, A.K., Saxena, N., Voris, J., Xiang, T.: Sensing-enabled defenses to RFID unauthorized reading and relay attacks without changing the usage model. In: Proceedings of the IEEE International Conference on Pervasive Computing and Communications, pp. 227–234 (2012)
18. Reid, J., Nieto, J.M.G., Tang, T., Senadji, B.: Detecting relay attacks with timing-based protocols. In: Proceedings of the 2nd ACM symposium on Information, Computer and Communications Security, pp. 204–213 (2007)
19. Kim, C.H., Avoine, G.: RFID distance bounding protocol with mixed challenges to prevent relay attacks. In: Garay, J.A., Miyaji, A., Otsuka, A. (eds.) CANS 2009. LNCS, vol. 5888, pp. 119–133. Springer, Heidelberg (2009). https://doi.org/10.1007/978-3-642-10433-6_9

20. Mitrokotsa, A., Dimitrakakis, C., Peris-Lopez, P., Hernandez-Castro, J.C.: Reid et al.'s distance bounding protocol and mafia fraud attacks over noisy channels. IEEE Commun. Lett. **14**(2), 121–123 (2010)
21. Ma, D., Prasad, A.K., Saxena, N., Xiang, T.: Location-aware and safer cards: enhancing RFID Security and privacy via location sensing. In: Proceedings of the ACM Conference on Wireless Network Security (WiSec), pp. 51–62 (2012)
22. Philipose, M., Smith, J.R., Jiang, B., Sundara-Rajan, K., Mamishev, A., Roy, S.: Battery-free wireless identification and sensing. IEEE Perv. Comput. **4**(1), 37–45 (2005)
23. Halevi, T., et al.: Sensing-enabled defenses to RFID unauthorized reading and relay attacks without changing the usage model. In: Proceedings of the IEEE International Conference on Pervasive Computing and Communications, pp. 227–234 (2012)
24. Czeskis, A., Koscher, K., Smith, J.R., Kohno, T.: RFIDs and secret handshakes: defending against ghost-and-leech attacks and unauthorized reads with context-aware communications. In: Proceedings of the ACM Conference on Computer and Communications Security (CCS), pp. 479–490 (2008)
25. Shu, Y., Gu, Y., Chen, J.: Sensor-data-enhanced authentication for RFID-based access control systems. In: Proceedings of the IEEE Mobile Ad Hoc Sensor Systems (MASS), pp. 236–244 (2012)
26. Cazecaa, M.J., Meada, J., Chenb, J., Nagarajana, R.: Passive wireless displacement sensor based on technology. Sens. Actuators A: Phys. **190**, 197–202 (2013)
27. Paggi, C., Occhiuzzi, C., Marrocco, G.: Sub-millimeter, displacement sensing by passive UHF RFID antennas. IEEE Trans. Antennas Propagation **62**, 905–912 (2014)
28. Tu, Y., Zhou, W., Piramuthu, S.: A novel means to address RFID tag/item separation in supply chains. Decis. Support Syst. **115**, 13–23 (2018)
29. Taghvaeeyan, S., Rajamani, R.: Nature-inspired position determination using inherent magnetic fields. Technology **2**(2), 161–170 (2014)
30. Avoine, G., Mauw, S., Trujillo-Rasua, R.: Comparing distance bounding protocols: a critical mission supported by decision theory. Comput. Commun. **67**(1), 92–102 (2015)
31. Gurulian, I., Akram, R.N., Markantonakis, K., Mayes, K.: Preventing relay attacks in mobile transactions using infrared light. In: Proceedings of the Symposium on Applied Computing, SAC 2017, pp. 1724–1731. ACM, New York (2017)
32. Gurulian, I., Markantonakis, K., Frank, E., Akram, R.: Good vibrations: artificial ambience-based relay attack detection. In: 2018 17th IEEE International Conference on Trust, Security and Privacy in Computing and Communications, pp. 481–489 (2018)
33. Tu, Y.-J., Piramuthu, S.: Lightweight non-distance-bounding means to address RFID relay attacks. Decis. Support Syst. **102**, 12–21 (2017)

Analysis of In-vehicle Security System of Smart Vehicles

Nazeeruddin Mohammad$^{(\boxtimes)}$, Shahabuddin Muhammad, and Eman Shaikh

Prince Mohammad Bin Fahd University, AlKhobar, Kingdom of Saudi Arabia
nmohammad@pmu.edu.sa

Abstract. Modern automobiles utilize numerous computerized systems to control and observe the state of the vehicle. These systems use in-vehicle networks to communicate with each other and make intelligent decisions. However, because of the increasing trend to offer connectivity between the in-vehicle network and external networks, automobiles are susceptible to several cyberattacks. This trend cannot continue unless the security of in-vehicle networks is enhanced to tackle all sorts of known and unknown attacks. This paper considers a multi-layered defense system for smart vehicles and models the in-vehicle security of the system. The model consists of a vehicle profile, communication profile, and considers several known defense mechanisms. The attacker profile is also captured based on its abilities and goals. We implement the proposed model in a probabilistic model checker and study the impact of a diverse attacker for a variety of defense profiles.

1 Introduction

Over the past two decades, advancements in the automobile industry remain unabated. Modern automobiles have now become an essential part of everyday life. Current modern vehicles consist of various controllers that are connected by different bus communication networks with various properties. The communication networks in automobiles have access to various components of the vehicle such as brakes, airbags, engine, and doors. Systems such as the electric power steering and anti-lock braking systems dismissed several of the conventional mechanical components which are replaced by sensors [3]. The in-vehicle network is developed to track these subsystems and enable communication among computers that observe several mechanical devices. A vast majority of Electronic Control Units (ECUs) are connected by several kinds of buses that control and track the situation of the vehicles. These ECUs use the in-vehicle network to communicate and control each other. Thus, in-vehicle communications decrease the complexity of designing, fixing and refitting in modern vehicles [11]. It also improves the safety, efficiency, and effectiveness of the intelligent vehicle systems [15].

Lately, the progress attained in communication networks, stereo vision, and computing has increased the interest of researchers in the field of in-vehicle

© Springer Nature Switzerland AG 2019
R. Doss et al. (Eds.): FNSS 2019, CCIS 1113, pp. 198–211, 2019.
https://doi.org/10.1007/978-3-030-34353-8_15

infotainment (IVI) systems. The IVI system enables the driver or vehicle to interact with other drivers or vehicles. This improves the overall traffic safety and quality of in-vehicle services. IVI consists of mainly two parts: the information part that facilitates safe driving, such as current traffic situations, advance warning notification, advice on important events, etc., and the entertainment part which is used to make the journey as amusing as possible, by providing TV shows, movies, advertisement, etc. Both the information and entertainment are combined to deliver a consolidated platform to the drivers and commuters. Even though vehicular infotainment systems are one of the essential components that automakers use to elevate their product's value, there are still some important threats that should be resolved before the vast adaption of IVI systems [7].

While in-vehicle communication networks guarantee safety against various technical interferences, they are mostly unprotected against malicious attacks and therefore bring up adversarial threats. As the demand for advanced vehicle systems rises, a great number of computer-controlled systems are being implemented into the vehicles by automobile industry manufactures. Since in-vehicle networks consist of several vulnerabilities, adversaries can easily attack a targeted vehicle as modern vehicles provide connectivity to the vehicle and the outside world. The environment in which modern vehicles transfer data is insecure and vulnerable to several attacks. If an adversary gains access to the network, the modern vehicle would have no power to stop them. Thus, the adversary can launch many attacks which could be dangerous or fatal to the car and driver [3]. To improve in-vehicle network security, it is crucial to understand the nature of the current attacks and propose appropriate solutions.

In this paper, we consider several defense mechanisms proposed in the literature and build an in-vehicle security framework. The proposed framework consists of a vehicle profile, communication profile, and defense profile. The framework also captures the abilities and goals of an attacker. The framework is modeled using the probabilistic model checking. The model is generic that can be adapted to study the in-vehicle security of modern vehicles. We implement the proposed model in the probabilistic model checker 'PRISM' and study the impact of a diverse attacker for a variety of defense profiles.

The rest of the paper is organized as follows. Section 2 briefly discusses the different ways to compromise in-vehicle security and their countermeasures. Section 3 describes the details of in-vehicle security and attacker models. Section 4 discusses the results obtained using the proposed model for different types of attackers and vehicle profiles. Finally, the paper is concluded in Sect. 5.

2   Literature Review

The smart vehicle platform is a common ground for attackers to destruct a vehicle or gain access to the in-vehicle network. Over the past few years, there have been numerous publications regarding security attacks against smart vehicles and their systems [6]. Recently, there have been significant technological advancements to enhance the convenience and entertainment in the vehicles.

To achieve these objectives, the vehicle infotainment system uses the Internet to entertain, support and inform the driver. Thus, automakers are integrating devices with common interfaces into their vehicles. This arises a possibility of vulnerabilities in the infotainment system that attackers can take advantage of for their benefit.

2.1   In-vehicle Attacks

In the following, we discuss the main ways in which in-vehicle security can be compromised.

Direct Attack: This attack occurs when external equipment is connected via network interfaces like CAN (Controller Area Network) bus. Any device that connects to the maintenance (OBD) port can take advantage of the bugs present in the in-vehicle systems and can control the vehicle's critical systems such as steering and braking [9]. The attacks on CAN bus can occur due to the broadcast nature of the bus and lack of message authentication and encryption. Due to the restricted computational power of ECUs, robust cryptographic algorithms are difficult to implement [2]. Further, to implement a robust data encryption algorithm, the CAN bus would need larger bandwidth than what is currently available [8].

External Devices: Automobile industry manufacturers are integrating a vast majority of the in-vehicle entertainment devices and applications. These external devices make the software applications present in the vehicle more accessible thereby introducing the possibility of applications containing malicious codes. Therefore, they have a high risk of attacks for the in-vehicle systems.

Outside Network Attacks: Smart keys, tire pressure monitoring system (TPMS), and inter-vehicular communications rely on short-range wireless communications that are susceptible to attacks such as eavesdropping. The pressure monitor device placed on the wheel valve consists of a 32-bit identifier that can interact remotely at a distance of around 40 m from passing vehicles. Its function is to present the pressure warning message given by the TPMS and then brighten up the alarm lamp instantly. Similarly, GPS has issues such as data tampering and release of privacy.

Embedded Web browsers: In-vehicle web browsers allow the completion of daily chores like emails, online payment, weather forecasts, and news updates on-board. Therefore, the in-vehicle networks are also susceptible to attacks faced by traditional computer-based networks. Embedded web browsers let the user access and download contents from the remote application stores on the Internet to the vehicles. These application stores might not be provided by the vehicle's automaker which may result in malware getting installed in the vehicles [17].

Removable Media Ports: The majority of modern vehicles are composed of USB ports for users to connect their external devices. Such interfaces enable the IVI system to gain access to the media files present on the external devices. Moreover, they are also used by the automakers to update the ECU firmware. The removable media can be infected with malicious malware that can extend into the vehicle's IVI system. Studies have mentioned a type of attack that is

launched due to the modified Windows Media Audio (WMA) file. This file, when played on the respective vehicle's media player, would send a malicious CAN message that would affect the IVI. Once a malware runs on a subsystem it can spread to the other parts of the system rendering the entire electronic equipment affected. Malware can also execute DoS attacks and inject fake messages to the in-vehicle networks to disturb other subsystems [17].

The above list is not exhaustive and there are several other attack methods reported in the literature. For instance, Costantino et al. [5] developed a social engineering attack named CANDY that contains a collection of remote attacks inflicted on the targeted automobile with an Android IVI system. The attacks are carried out by exploiting the android app developed and designed by the authors. The app acts as a trojan horse and exposes a backdoor that enables the hacker to remotely access the information present on the CAN bus and personal information of the driver. Similarly, in [12] weaknesses in the MirrorLink protocol present in the modern vehicle is demonstrated. The authors developed a malicious app that takes advantage of the vulnerabilities present in the MirrorLink protocol. It allows the hacker to access a privileged process executing on the IVI. Moreover, the hacker can also attack the smartphone of the driver and send malicious messages to the internal CAN bus of the vehicle.

2.2 Attack Countermeasures

In the following, we discuss some in-vehicle attack countermeasures proposed in the literature.

A security framework was proposed for vehicular systems to overcome the message authentication problem of the CAN bus in [14]. The proposed method is validated by implementing a prototype on a testbed using the Fessscale automotive development board, in which every node that sends a CAN packet is required to send a Message Authentication Code (MAC) packet of size 8 bytes. The ECUs were subdivided into two categories known as the low-trust group and high-trust group. The low-trust group consists of external interfaces and telematics, whereas the high trust-group contains ECUs that have no external interfaces. Moreover, the high-trust group shares a secret key to validate and verify all incoming and outgoing messages, and ECUs from the low-trust group cannot transmit messages to critical ECUs in the high-trust group. This method uses SHA-3 hash function with an improvement in the system throughput by pre-calculation of the computationally intensive cryptographic function, which ensures 2000 extra clock cycles as compared to the system with no MAC method.

Security software is proposed for CAN which uses CANoe and the load on the bus was measured using a CANcaseXL device [16]. The messages on CAN bus are compressed by using a CAN data compression algorithm (ECANDC). AES-128 is used to encrypt the data and HMAC is used for authentication. The method is implemented on 20 ECUs and it is concluded that this method can be used for in-vehicle networks. Encryption and authentication of the respective CAN messages are achieved with an average message delay which is lower than 0.13 ms.

Kang et al. [10] proposed a novel deep neural network (DNN) based on intrusion detection (IDS) system to enhance the security of the in-vehicular network. A deep neural network was utilized to provide the chances for each class to differentiate normal packets and hacking packets which would make the system be able to recognize any malicious attack that occurs to the automobile. Moreover, the authors also proposed a feature vector containing the nature and the value of the information taken from the network packet. The accuracy achieved by the system is 98%.

In [1], the authors proposed D2H-IDS which is an IDS that can be used in smart vehicles. The system is used to prevent security attacks and is implemented through a three-step data traffic search, reduction, and grouping method that is used to separate trusted service requests against fake requests that are generated during intrusion attacks. The accuracy accomplished with the system was 99.43% with a 99.92% detection rate.

Cho et al. [4] proposed an IDS named Clock-based IDS (CIDS) for the protection of ECUs against in-vehicle network attacks. The proposed system was used to recognize various types of in-vehicle network attacks on a CAN bus prototype as well as on an automobile. The success rate of identifying in-vehicle network intrusions has received a low false-positive rate of 0.055% and a high true positive rate of 100%.

Ben et al. [13] developed a novel method to identify the "no attack" and "under-attack" states of the connected vehicles. K-means clustering, Hidden Markov Model and Pearson correlation were used for the detection of injected fabricated message attacks. No attack means that there are no injection of speed reading and Revolutions Per Minute (RPM) messages whereas under-attack means that the forged speed reading or RPM reading messages are inserted.

3   In-vehicle Security Modeling

Modern vehicles are offering increased connectivity through various wired and wireless interfaces. These interfaces can be exploited by adversaries to launch attacks on various internal systems. In particular, using IVI systems the remote attacks could be easily launched. Vehicle manufacturers are implementing diverse countermeasures to tackle security issues. The main countermeasures include the partition of external interfaces from the in-vehicle network, incorporation of encryption and authentication for in-vehicle traffic, and installation IDS for anomaly detection. In this section, the influence of these defense mechanisms against various in-vehicle attacks is modeled using probabilistic model checking.

For this study, a comprehensive multi-layered defense framework is modeled. Depending on the vehicle manufacturer's preference one or all of them are implemented. The first level of defense is implemented in apps or user interfaces that vehicle offers to the users. The second level of defense like a firewall or a gateway, which is installed between the external interfaces and the internal network to filter the traffic. The data that is not supposed to leave the internal network will be blocked. Similarly, any unwanted/uncommon traffic originating

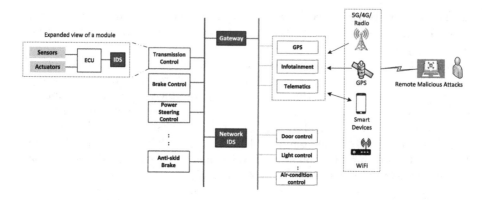

Fig. 1. A schematic diagram of in-vehicle network with defense mechanisms.

from external interfaces is also hampered. Finally, in-vehicle IDS can be deployed to detect known and unknown attacks. IDS can be placed at two locations: on ERUs or the in-vehicle network. IDS deployed at each critical ERU is fine-tuned to detect anomalies related to that ERU. This type of deployment helps detect the origin of the attack as well. However, IDS deployed at ERU increases the complexity of the system considerably. On the other hand, an IDS placed on the in-vehicle network can detect the anomalies on the communication bus, and doesn't burden the resource-constrained ERUs. A schematic diagram of an in-vehicle network along with defense mechanism is shown in Fig. 1.

A vehicle can be attacked via physical (direct wired) access or wirelessly. The direct wired attack is carried out through an OBD port present in the vehicle whereas a wireless attack can be done via various technologies. Low range wireless attacks can be done from within the vicinity of the vehicle using Bluetooth or Wi-Fi technology. The long-range attacks can be done using the smart devices which are paired with IVI. A Smart App which is provided for this purpose by the vendor can be compromised or another malicious App can be installed on this phone.

The vehicle's security system comprises of several components that consist of accessibility vector, vulnerability vector, and defense profile of the vehicle. The accessibility vector of a vehicle defines the means and the associated score value with which the vehicle can be attacked. For example, if a vehicle has an OBD port and Bluetooth interface then its accessibility vector contains the associated cost of these two parameters. The vulnerability vector of a vehicle defines the probability values of an attack being successful given the accessibility vector. The vehicle may implement multiple levels of defense such as gateway segregating the low and high priority buses, intrusion detection systems (IDS) at the network and host (ECU) levels, and authenticated communication and encrypted CAN messages.

In addition to the vehicle's security system, the attacker is also modeled based on its abilities and intentions. The attacker's profile captures the parameters such

as the type of attack (wired or wireless), the attacker's vicinity (low range or high range), and whether the attacker's intention to launch the attack is of low or high impact.

Once the aforementioned parameters are defined, the model of the system M can be checked against a variety of properties of interest ϕ. The model checker either satisfies the specification ($M \models \phi$) or produces a counterexample. It can also be used to calculate the probability or reward of reaching a state of the system. In particular, after implementing the system's model in PRISM language, we use linear-time temporal logic (LTL) as the property specifications language. For instance, the following LTL formula calculates the probability that an attack will become successful in a given time.

$$P =? \ [F \leq Time \ \ AttackSuccess]$$

Here F is the temporal operator that represents "future" states of the system and $AttackSuccess$ a user-defined formula that represents what constitutes a successful attack. To calculate the reward accumulated over a set of states or a path of the model, a reward structure is used. For example, the following reward structure captures the cost of an attack:

    rewards "attack_cost"
        [attack_type_1] condition_1 : cost_1;
        [attack_type_2] condition_2 : cost_2;
        :
        [attack_type_n] condition_n : cost_n;
    endrewards

Here 'condition_i' is a boolean formula that becomes true when 'attack_type_i' is launched. Similar to the LTL formula given above, the following property specification can be used to calculate the cost of a successful attack:

$$R\{\text{"}attack\_cost\text{"}\} =? \ [F \leq Time \ \ AttackSuccess]$$

PRISM model checker can check the given property specification against a system model. Its output is either to validate that the system meets given specifications or to find the related probability/reward. An example output is shown in Fig. 2. The working details of the proposed in-vehicle security system are described in Algorithm 1.

Property:
R{"attack_cost"}min=? [C<=T]

Defined constants:
p_wired=0,p_li=0,p_lr=0.5,pa_obd=0.8,pa_bt=0.5,pa_wf=0.5,pa_a1=0.5,pa_a2=0.5,pd_gw=0.5,pd_idsn=0.5,pd_ae=0.5,pd_idsh=0.9, T=300

Method:
Verification

Result (minimum expected attack_cost):
10.887987426754767 (value in the initial state)

Fig. 2. The sample output from PRISM model checker.

Algorithm 1. The in-vehicle security system

1: **procedure** SYSTEM
2:    Init: $t \leftarrow 0$, $S \leftarrow \emptyset$, $comp = false$, $Succ_a = false$, $Succ_d = false$
3:    Init: A_a, A_v, P_d ▷ accessibility vector, vulnerability vector, defense profile
4:    Init: $P_a \in \{P_1, P_2, P_3\}$ ▷ attacker's profile $P_i = (Comm, Imp, Range)$
5:    Define: $\Phi = \bigvee\limits_{i \in D_p} \phi_i$ ▷ attack vector for $D_p = f(P_a)$, $\phi_i = \top$ when $d_i \in D_p$ is
compromised
6:    **while** *true* **do**   ▷ run for all the components of selected attack profiles
7:       select $c \in Comm$, $i \in Imp$, $r \in Range$ ▷ non-deterministic selection
8:       select $d \in D_p$, $(d \notin S)$ ▷ non-deterministic selection
9:       $t \leftarrow t + 1$
10:       **if** $d = \emptyset$ **then** break
11:       select $Succ_a \leftarrow true$ with prob. p_a^d ▷ attack success on component d
12:                  $\leftarrow false$ with prob. $(1 - p_a^d)$, $S = S \bigcup d$
13:       **if** $Succ_a = false$ **then go to** 8
14:       **if** $P_a(Imp) = low$ **then**
15:          select $Succ_d \leftarrow true$ with prob. p_d^{AE}, $S = S \bigcup d$ ▷ defense type AE
successful
16:                     $\leftarrow false$ with prob. $(1 - p_d^{AE})$, $comp = true$, break
17:       **else**
18:          select $Succ_d \leftarrow true$ with prob. p_d^{GW}, $S = S \bigcup d$ ▷ defense type GW
successful
19:                     $\leftarrow false$ with prob. $(1 - p_d^{GW})$
20:          **if** $Succ_d = true$ **then go to** 8
21:          select $Succ_d \leftarrow true$ with prob. $p_d^{IDS_n}$, $S = S \bigcup d$ ▷ defense type IDS_n
successful
22:                     $\leftarrow false$ with prob. $(1 - p_d^{IDS_n})$
23:          **if** $Succ_d = true$ **then go to** 8
24:          select $Succ_d \leftarrow true$ with prob. p_d^{AE}, $S = S \bigcup d$ ▷ defense type AE
successful
25:                     $\leftarrow false$ with prob. $(1 - p_d^{AE})$
26:          **if** $Succ_d = true$ **then go to** 8
27:          select $Succ_d \leftarrow true$ with prob. $p_d^{IDS_h}$, $S = S \bigcup d$ ▷ defense type IDS_h
successful
28:                     $\leftarrow false$ with prob. $(1 - p_d^{IDSH})$
29:          **if** $Succ_d = true$ **then go to** 8
30:          **else** $comp = true$, break

4   Implementation and Analysis

The model presented in the previous section is implemented in a well-known probabilistic model checker PRISM. The implementation is used to assess the impact of different types of attackers. We consider the following three attacker's profiles in this paper:

1. Profile I: The attacker prefers to perform high impact attack using direct access to the vehicle such as OBD interface. This profile corresponds to

the following probabilities: $p_{direct} = 90\%$, $p_{wireless} = 10\%$, $p_{highImpact} = 100\%$. Moreover, the attacker chooses the wireless attacks of short range as well as long range with equal probability $p_{lowRange} = p_{highRange} = 5\%$

2. Profile II: In this profile, the attacker performs low impact attack and prefers low range wireless communication medium. The corresponding probabilities are: $p_{direct} = 10\%$, $p_{wireless} = 90\%$, $p_{lowImpact} = 100\%$, $p_{lowRange} = 81\%$

3. Profile III: The attacker wants to do high impact attacks and prefers long range wireless communication medium. The corresponding probabilities are: $p_{direct} = 10\%$, $p_{wireless} = 90\%$, $p_{highImpact} = 100\%$, $p_{lowRange} = 9\%$

The attacker tries to launch an in-vehicle attack on a vehicle that has the following characteristics:

1. Accessibility Vector (A_a): This captures the cost of launching an attack on a vehicle where it is relatively difficult to access OBD port as compared to wireless channels. The corresponding accessibility vector is captured by the following values: $A_{OBD} = 10$, $A_{BT} = 2$, $A_{WF} = 1$, $A_{DA} = 3$, $A_{IA} = 5$.

2. Vulnerability Vector (A_v): It captures the probability of successfully exploiting a vulnerability. Once the OBD port becomes accessible, the chance of launching an attack is higher than attacking the vehicle through any other means. Therefore, we choose a higher success probability for OBD (80%) and keep the remaining probabilities at 50%.

3. Defense Profile (P_d): The defense profile of a vehicle captures the probability of success for each defense mechanism. It is a vector of four probabilities corresponding to four defense mechanisms GW, IDS_N, AE, and IDS_H. We assume vehicles with four defense profiles where the defense are incrementally added as given by the following vectors:
   VP$_1$=[0.5, 0, 0, 0]
   VP$_2$=[0.5, 0.5, 0, 0]
   VP$_3$=[0.5, 0.5, 0.5, 0]
   VP$_4$=[0.5, 0.5, 0.5, 0.9]

In the following, we present the probability of attack success and the attacker's cost for the aforementioned attacker and vehicle profiles.

Figure 3 shows the probability of attack success for four vehicle profiles. It can be seen from this figure that with the addition of each defense mechanism, the probability of launching a successful attack reduces such that the overall probability of attack being successful reduces down to 4% when all the defense mechanisms are in place. The corresponding increase in the attack cost is shown in Fig. 4. With every addition of defense mechanism, the attacker incurs more attempts in order to launch a successful attack.

Figures 5 and 6 show the corresponding probability of attack success and attack cost for attacker profile II. Since the attacker is only interested in the low impact attacks, the attacks can be done without any involvement of the critical components on the high-speed bus. That means, the defense mechanism at the gateway as well as the influence of the IDS at the network and the host have no impact in this scenario. The only defense mechanism that is trying to thwart the

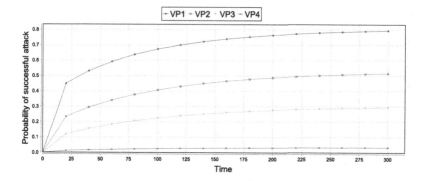

Fig. 3. The probability of attack success for profile I.

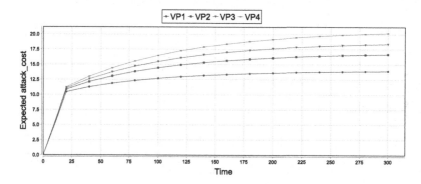

Fig. 4. The attack cost for profile I.

attack is AE which results in two behaviors shown in the figures. The vehicle profile VP_1 and VP_2 have the same probabilities of attack success. Due to the presence of AE in VP_3 and VP_4, they show the same behavior. This is reflected in Figs. 5 and 6. It can also be noted that the probability of a successful attack is drastically increased in this attacker's profile which means that the attacker is highly likely to launch the successful attack. This provides us insight that the low impact attacks can be reduced by using higher authentication standards or by including IDS in the non-critical ECUs as well. However, it will require higher processing at ECUs.

Figures 7 and 8 show the probability of a successful attack and the corresponding cost for the attacker's profile III. Since the attacker is inclined to do high impact attack relying mostly on wireless links, it has multiple avenues to exploit and launch an attack such as via Bluetooth, WiFi, etc. Because of this reason, the probability that the attacker will be able to successfully launch an attack increases quickly as compared to attacker's profile I. In other words, the attacker can quickly launch an attack if it can exploit any vulnerability either in the wireless medium or in the application interface.

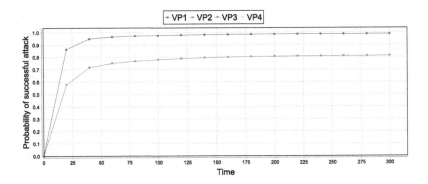

Fig. 5. The probability of attack success for profile II.

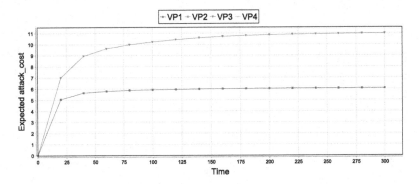

Fig. 6. The attack cost for profile II.

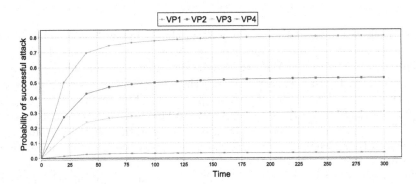

Fig. 7. The probability of attack success for profile III.

Although each defense mechanism plays its role in increasing the security of the in-vehicle security system, the effect of AE and IDS_h needs further elaboration especially when the impact of the attack is under consideration. For low impact attacks, AE is the only defense mechanism and thus other

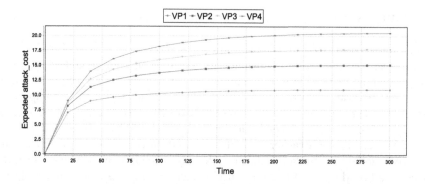

Fig. 8. The attack cost for profile III.

defense mechanisms don't play any role in the probability of attack success or failure. However, when the attack is generic as the attacker tries to launch both low impact as well as high impact attacks, each defense mechanism plays its role and tries to thwart the attack in addition to AE. This is demonstrated in Fig. 9. Moreover, when the attacker chooses solely high impact attack, then not only AE but other defense mechanisms start playing important roles with IDS_h appearing to be the dominant defense line. This is shown in Fig. 10.

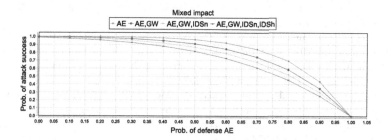

Fig. 9. The probability of mixed impact (both HI and LI) attack success vs strength of AE.

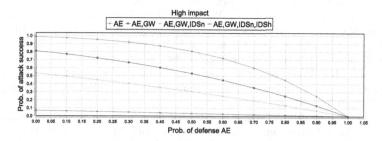

Fig. 10. The probability of high impact attack success vs strength of AE.

5   Conclusion

Several research studies in recent years have shown the practicability of the cyberattacks on the modern automobiles because of their connectivity with the external networks. This is becoming more concerning as the adversaries can launch cyberattacks through wireless in-vehicle networks. In this paper, we have modeled in-vehicle security using probabilistic model checking. A vehicle is modeled based on its accessibility vector, vulnerability vector and defense profile. Moreover, the attacker is modeled based on its goals and abilities. The presented model is generic, which can be fine-tuned to capture the security aspects of a broad range of manufactured vehicles and assess their vulnerabilities against different types of attacks. We implemented the model in the PRISM model checker and studied the results for three diverse attacker profiles and four distinct defense profiles. The results demonstrated the impact of each defense mechanism, especially the role of authentication and encryption in securing the vehicle. As part of our future work, we intend to extend the model and include various protocols and platform vulnerabilities in smart vehicles by major vehicle manufacturers.

References

1. Aloqaily, M., Otoum, S., Al Ridhawi, I., Jararweh, Y.: An intrusion detection system for connected vehicles in smart cities. Ad Hoc Netw. **90**, 101842 (2019). https://doi.org/10.1016/j.adhoc.2019.02.001
2. Buttigieg, R., Farrugia, M., Meli, C.: Security issues in controller area networks in automobiles. In: 2017 18th International Conference on Sciences and Techniques of Automatic Control and Computer Engineering (STA), pp. 93–98. IEEE (2017). https://doi.org/10.1109/STA.2017.8314877
3. Carsten, P., Andel, T.R., Yampolskiy, M., McDonald, J.T.: In-vehicle networks: attacks, vulnerabilities, and proposed solutions. In: Proceedings of the 10th Annual Cyber and Information Security Research Conference, p. 1. ACM (2015). https://doi.org/10.1145/2746266.2746267
4. Cho, K.T., Shin, K.G.: Fingerprinting electronic control units for vehicle intrusion detection. In: 25th {USENIX} Security Symposium ({USENIX} Security 16), pp. 911–927 (2016). https://dl.acm.org/citation.cfm?id=3241165
5. Costantino, G., La Marra, A., Martinelli, F., Matteucci, I.: Candy: a social engineering attack to leak information from infotainment system. In: 2018 IEEE 87th Vehicular Technology Conference (VTC Spring), pp. 1–5. IEEE (2018). https://doi.org/10.1109/VTCSpring.2018.8417879
6. Fysarakis, K., Askoxylakis, I., Katos, V., Ioannidis, S., Marinos, L.: Security Concerns in Co-operative Intelligent Transportation Systems (2017). https://doi.org/10.1201/b21885-16
7. Guo, J., Song, B., He, Y., Yu, F.R., Sookhak, M.: A survey on compressed sensing in vehicular infotainment systems. IEEE Commun. Surv. Tutorials **19**(4), 2662–2680 (2017). https://doi.org/10.1109/COMST.2017.2705027
8. Hartzell, S., Stubel, C.: Automobile CAN bus network security and vulnerabilities (2018). https://canvas.uw.edu/files/47669787/download?download_frd=1

9. Huang, Y., He Qin, G., Liu, T., Dan Wang, X.: Strategy for ensuring in-vehicle infotainment security. Appl. Mech. Mater. **556–562**, 5460–5465 (2014). https://doi.org/10.4028/www.scientific.net/AMM.556-562.5460

10. Kang, M.J., Kang, J.W.: Intrusion detection system using deep neural network for in-vehicle network security. PloS One **11**(6), e0155781 (2016). https://doi.org/10.1371/journal.pone.0155781

11. Liu, J., Zhang, S., Sun, W., Shi, Y.: In-vehicle network attacks and countermeasures: challenges and future directions. IEEE Netw. **31**(5), 50–58 (2017). https://doi.org/10.1109/MNET.2017.1600257

12. Mazloom, S., Rezaeirad, M., Hunter, A., McCoy, D.: A security analysis of an in-vehicle infotainment and app platform. In: 10th {USENIX} Workshop on Offensive Technologies ({WOOT} 16) (2016). https://www.usenix.org/conference/woot16/workshop-program/presentation/mazloom

13. ben Othmane, L., Dhulipala, L., Abdelkhalek, M., Govindarasu, M., Multari, N.: Detection of injection attacks in in-vehicle networks (2019). https://lib.dr.iastate.edu/ece_conf/72

14. Wang, Q., Sawhney, S.: Vecure: a practical security framework to protect the CAN bus of vehicles. In: 2014 International Conference on the Internet of Things (IoT), pp. 13–18 (2014). https://doi.org/10.1109/IOT.2014.7030108

15. Willke, T.L., Tientrakool, P., Maxemchuk, N.F.: A survey of inter-vehicle communication protocols and their applications. IEEE Commun. Surv. Tutorials **11**(2), 3–20 (2009). https://doi.org/10.1109/SURV.2009.090202

16. Wu, Y., Kim, Y.-J., Piao, Z., Chung, J., Kim, Y.-E.: Security protocol for controller area network using ECANDC compression algorithm. In: 2016 IEEE International Conference on Signal Processing, Communications and Computing (ICSPCC), pp. 1–4 (2016). https://doi.org/10.1109/ICSPCC.2016.7753631

17. Zhang, T., Antunes, H., Aggarwal, S.: Defending connected vehicles against malware: challenges and a solution framework. IEEE Internet Things J. **1**(1), 10–21 (2014). https://doi.org/10.1109/JIOT.2014.2302386

Quorum Chain-Based Malware Detection in Android Smart Devices

Fei Gao[1], Frank Jiang[2(✉)], Yuping Zhang[3], and Robin Doss[2]

[1] College of Electronics Engineering, Guangxi Normal University, Guilin, China
[2] School of Info Technology, Deakin University, Geelong, Australia
FrankNSW2011@gmail.com
[3] Chengdu Technological University, Chengdu, China

Abstract. Smart devices are gradually becoming indispensable in people's daily lives, and Android-based smart devices are taking over the main stream in mobile devices. However, while Android smart devices bring convenience to customers, they also bring problems. Due to the open-sourced nature of the Android system, malicious programs and software attacks pose a significant security risk to user data. Therefore, the detection of malware has always been a critical issue. For a long time, various malware detection schemes have been proposed, which have gradually improved the detection of malware. Traditional detection methods are based on static or dynamic detection techniques. In recent years, with the advancement of technology, malware detection based on machine learning ideas has been widely used, such as K-NN, deep learning, decision trees, and so on. Blockchain has been widely used in many fields since its birth. This paper combines traditional detection methods and ensemble learning algorithms to propose a malware detection technology based on QuorumChain framework (blockchain technology). The experimental results verify that the proposed new model is better than other models in precision, recall and f1-measure.

Keywords: Android devices · Malware detection · Quorum chain

1 Introduction

With the advancement of mobile technology, the smart phones are gradually taken over partial functions of personal computers in people's daily life. According to the statistics of 2018, the current mobile smart device market, Android and IOS systems accounted for more than 99%, of which IOS devices accounted for about 14.5%, Android devices accounted for more than 85% [1]. The open-sourced nature of the Android system makes it vulnerable to many cyber-attacks, such as malwares. Mobile malware refers to software that typically has malicious code for applications. Install or run a module or code fragment software [2] that violates relevant national laws and regulations on the mobile terminal to achieve illegal purposes. The emergence of malicious applications poses a serious threat to user data. The types of malicious applications include malicious deduction, stealing privacy, remote control, malicious transmission, system destruction, fraud, etc., which can lead to the loss of user privacy data and cause economic losses to users.

© Springer Nature Switzerland AG 2019
R. Doss et al. (Eds.): FNSS 2019, CCIS 1113, pp. 212–224, 2019.
https://doi.org/10.1007/978-3-030-34353-8_16

Currently, the detection of smart-phone malwares become more and more crucial. Detection techniques mainly include static analysis [3, 4] and dynamic analysis [5]. Static analysis is based on the static syntax and function structure of the program to determine potentially malicious code. This method has a high code coverage, but because there is no actual program running, the false alarm rate is higher; the manufacturer will use techniques such as code obfuscation to avoid static analysis. Dynamic analysis refers to running an application in a simulated environment. During the operation, malicious behaviour is identified based on the acquired behaviour information. All discovered are real malicious applications. The drawback is low code coverage and high overhead. Traditional static signature detection methods do not recognize confusing malicious code. The accuracy of software code detection for mutations is not very high, and dynamic detection methods require a large amount of resources, so most detection systems are now based on dynamic and static detection techniques.

2 Related Work

With the popularity of mobile devices, people's lives are inseparable from mobile devices. Therefore, the security of mobile devices has always been the focus of attention. Therefore, the security of mobile devices has always been the focus of attention, and personal security issues have also been severely challenged. In terms of the protection of mobile devices, from theory to application, the core is static analysis and dynamic analysis.

2.1 Malware Detection Scheme Based on Static Analysis

Di Cerbo et al. [6] proposed to identify sensitive licenses from the application, use Apriori algorithm to mine the correlation to form a malware library, and realize the identification of unknown software. Fuchs et al. [7] proposes a data stream static analysis technology that can automatically infer the security of Android applications using the ScanDroid tool. Shang, Li, Deng et al. [8] proposes a naive Bayesian classification algorithm based on machine learning is proposed to classify and detect software permissions. Use the posterior probability to calculate the most likely classification target value to achieve software detection.

2.2 Malware Detection Scheme Based on Dynamic Analysis

Li, Shen, Sun et al. [9] proposed the DroidAddMiner system for detecting and classifying Android malware features. The detection accuracy of malware is 98%, and the false positive rate is 0.3%.

Bhatia and Kaushal [10] proposed a system call capture system, SysCall-Capture, to track application calls to the system. The main function of the calling function set is to run the software on the virtual machine. A software feature data set is formed, and the data set is processed by the decision tree and the random forest to complete the classification of the application.

Enck and Gilbert [11] designed a dynamic detection tool for TaintDroid. The tool smudges a variety of sensitive data and then monitors the flow path of these contaminated sensitive data in real time in a sandbox environment to determine whether the software has malicious behaviour of privacy breaches.

2.3   Combination of Dynamic and Static Detection Scheme

Both dynamic and static analysis have their own advantages and disadvantages. The combination of the two will greatly improve the detection of malware.

The AASandbox [12] system is the first system proposed to combine static and dynamic analysis on the Android platform. Static analysis, decompile program files to identify malware behaviour. Dynamic analysis, using Android Simulator and Monkey Tools to simulate user behaviour and analyse simulator log files to identify malware.

Su and Fung et al. [13] proposed a combination of dynamic and static analysis. Divided into two levels, the static analysis layer is used to extract application permissions to form a vector set. The dynamic analysis layer uses sandbox analysis software to run log files to identify hidden malicious behaviour.

The use of dynamic and static combined detection methods improves the accuracy of software detection and reduces the cost of time. The detection of malware based on dynamic and static detection schemes has been continuously proposed, and the test accuracy has been greatly improved. In the development of detection technologies, we have also seen constant technological updates, such as deep-learning-based malware detection [14], Markov-based malware detection [15], etc.

2.4   Blockchain and QuorumChain

In 2008, with the launch of virtual currency Bitcoin [16], it received a lot of attention. As the core technology of Bitcoin, the blockchain has also entered the public's field of vision. Blockchain [17] is a distributed system that is highly secure due to its decentralized [18], non-destructive and traceability features.

Blockchain technology has been used widely in finance, healthcare, media, Internet and IoT etc. Most of the current preservation of social data is on the central node, and data access must pass through the central organization. This centralized organization has certain security issues in terms of data preservation. When the protection of a centralized organization is compromised, the data stored in the organization is lost and stolen, which makes the security of the data very threatened. Blockchain technology makes up for these shortcomings, and the decentralized public chain and multi-centered federal chain can prevent data loss. The data stored in each node is equivalent and does not cause data loss due to data loss of one node. The inextricable modification and traceability of the blockchain makes it possible to track data access and play an important role in financial transactions and network access.

QuorumChain is an enterprise-based distributed ledger platform stemmed from Ethereum's blockchain network. Data transfer between QuorumChain nodes adds a privileged function to the original unrestricted P2P transport mode, so it can only be

transmitted between nodes that allow each other. QuorumChain is a consensus algorithm based on voting mechanisms. QuorumChain is different from the common blockchain data sharing mechanism. The common blockchain shares its data to all nodes in the network, and all users jointly record and maintain data security. QuorumChain has a common block sharing mechanism for public data, and a private data transfer mechanism has been added. Its state storage is divided into two parts: public state and private state. The public state holds the consensus of the entire node and can be viewed in all nodes. The private state only records information about private transactions, and transactions are only allowed to be viewed at nodes that reach a private consensus. Because of these advantages, we can achieve data point-to-point transmission based on the consortium chain.

3   Contribution

In this paper, the major contributions are two-fold:

(1) A new model/approach in combination of Ensemble learning [19] and QuorumChain is proposed and validated. Ensemble learning is a machine learning method that uses a series of learners to learn and uses a certain rule to ensemble individual learning results to achieve better learning outcomes than a single learner. QuorumChain is based on Ethereum but has its own characteristics. The data of each node of QuorumChain is not completely the same, and QuorumChain adds mutual authorization private transmission mechanism between nodes in data transmission. So that different data can be processed on each node and the main features of the data set are finally obtained.

(2) Dataset contribution: Combined with ensemble learning and QuorumChain, the input dataset is characterized by the base classifier of each node. The voting mechanism is then used to filter the features acquired by the base classifier to obtain the salient features of the software.

3.1   Model Overview

As shown in Fig. 1, combining the characteristics of the QuorumChain and ensemble learning [20], we can apply the same or different basic classifiers on each node of QuorumChain. The input data is processed by the basic classifier of each node, and the feature is filtered by the voting mechanism to obtain the optimal data. Firstly, malware samples and benign software samples are obtained from the network, and static and dynamic feature extraction is performed on the samples in the sample set to form a software feature set. The feature set is preliminarily screened, and the filtered feature set is selected multiple times in a put-back manner. Then input to the QuorumChain nodes for main feature extraction. Finally, the importance of each feature is determined by voting. The main characteristics of the data set.

Figure 2 shows the model of malware detection developed in this paper, which mainly includes feature extraction module and classification detection module. The software sample set for training (including malware and benign software) extracts the static and dynamic features of the software through the feature extraction module and quantizes the features to obtain the data set. These features are sampled, and the bagging has a data set with M sample features selected. The N data sets are input to the classification detection module, and the input sample set is processed by the weak separator. Finally, the feature library of the training sample set is obtained by voting, including the malware signature database and the benign software signature database.

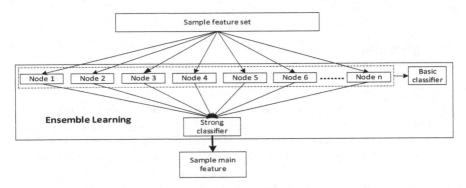

Fig. 1. Ensemble learning framework

3.2   Android Software Feature Extraction Model

The features of Android software mainly include static features (such as Android application permissions, Sensitive API calls, etc.) and dynamic features (such as file reading, network communication, application process operation, etc.), so the feature extraction module includes static feature extraction and dynamic feature extraction.

(1) Static Feature Extraction
    (a) Decompose the APK using APKTool [21] to get the AndroidManifest.xml file and the bytecode file (class.dex);
    (b) Use python to extract software functionality from AndroidManifest.xml files and bytecode files, and view permissions in files, monitor system events, permission rates, and sensitive API information;
    (c) Quantize the extracted software features. If the feature exists as "1", otherwise it is marked as "0" to form a static feature vector. Each feature vector represents a sample, and all feature vectors are combined into an experimental data set (Fig. 3).

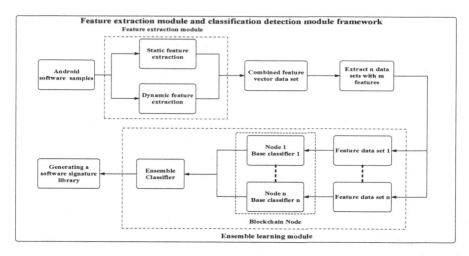

Fig. 2. Feature extraction process

(2) Dynamic Feature Extraction

    (a) Obtain the AndroidManifest.xml file of the APK installation package by pre-processing, extract the key information such as the application package name, component information, application permission information and main Activity name in the file, as the input of the next dynamic detection;

    (b) Start the virtual machine emulator and install the application in the emulator. After the installation is complete, run the program and perform dynamic detection;

    (c) Run the trigger sub-module, extract and process the control layout elements of the application, generate the corresponding trigger strategy and execute, and finally get the software behavior detection result. After the dynamic detection is completed, the feedback test result is stored to obtain the corresponding data table of the software;

    (d) Analyze the dataset to obtain the dynamic characteristics of the software and quantify the dynamic features. The feature representation is denoted as "1", otherwise it is denoted as "0", constituting a dynamic feature vector. Each feature vector represents a sample, and all feature vectors are combined into an experimental data set.

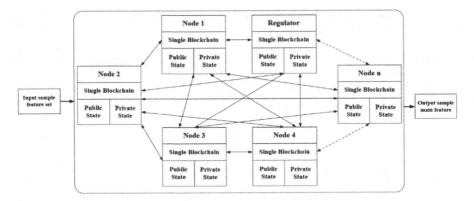

Fig. 3. QuorumChain-based classification detection model

3.3   Classification Model

QuorumChain is based on Ethereum, a consensus-based voting algorithm that is implemented through smart contracts and consensus. Through the smart contract structure of Ethereum, when the QuorumChain network is created, the voting smart contract will be compiled and stored in the creation block, and then the voting algorithm will be shared by each node through the QuorumChain network.

Here, the feature set is filtered by the ensemble learning idea. The set of feature vectors extracted from the sample is randomly divided into sub-vector sets, sent to each node and weakly classified (decision tree) to obtain sample features. Then the classification results obtained by each node are voted to form a strong classifier (random forest). Finally, the results of each node are voted on node N, and the software feature library of software feature generation is output. Random forest strong classifier generation steps:

(a) The bootstrapping method is used to extract m samples from the original training set, and n extractions were performed to generate n_tree training sets;
(b) For n_tree training sets, generate n decision tree models through training;
(c) For a single decision tree model, the selection of split nodes is performed according to the selection of the smallest Gini index to construct the decision tree.
(d) Generate multiple decision trees to form a random forest and determine the final prediction result by voting.

4   Detection Process

As shown in Fig. 4, the sample to be tested obtains a feature set of the software through a software feature extraction model. Then the software feature set is input into the trained random forest model, the software is analysed through the decision tree, and

finally the software is determined to be malware or benign software through the voting mechanism.

As shown in Fig. 5, the smart contract performs correlation validation on the input software features/characteristics to determine whether the software is malware, which is at the heart of software detection. As can be seen from Fig. 5, the malware chain and the benign software chain exist simultaneously with each node, and the malicious feature library and the benign feature library are respectively stored. When the software to be tested forms a feature vector set after feature extraction, the extracted software features are compared with the feature database stored in the node through the smart contract to determine whether the tested software is malware or benign software. In this way, a new block is formed in the node, and the block is added to the corresponding block chain (malware chain/benign software chain), which further enriches the feature library.

The aforementioned depicts the main detection process for software malware, combined with dynamic and static detection models. Considering the accuracy of the model for the detection, we calculate the accuracy rate, recall rate and F1-measure of the test results:

$$Recall = \frac{TP}{TP + FP} \quad \Pr ecision = \frac{TP + TN}{TP + FP + TN + FN}$$

$$F1\text{-}measure = \frac{\Pr ecision \cdot \text{Re}call}{\Pr ecision + \text{Re}call}$$

where TP represents a true class, FP represents a false positive class, TN represents a true negative class, and FN represents a false negative class.

5   Analysis

For the detection of malware, its accuracy and recall rate are the criteria for measuring the pros and cons of the test program. There are 1000 samples in this experiment, of which malicious samples and benign samples each account for half. Malicious sample gets from https://iscxdownloads.cs.unb.ca/iscxdownloads/CICAndMal2017/APKs/, Benign samples were obtained from the Android app market. There are four types of malicious samples obtained: 1. Adware; 2. Ransomware; 3. SMS malware; 4. Scareware. We extract the static features and dynamic features of 1000 samples, extract the static features of the samples through de-compilation techniques, and obtain the dynamic characteristics of the software using the sandbox environment. We use the characteristics of each sample as a vector to separate the malware(s) and the benign software in order to get two vector groups.

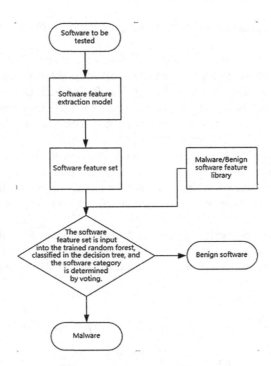

Fig. 4. Flowchart of detection model

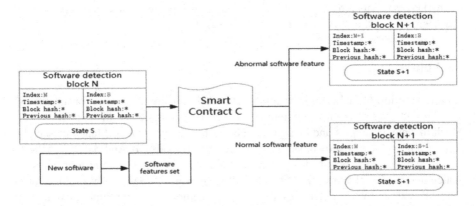

Fig. 5. Core detection model

Firstly, the software feature vector group is pre-processed and the features are initially screened. Then, the feature vector set is filtered by ensemble learning algorithm to obtain the salient features of the software. We randomly selected 300 malware and 300 benign software from the sample as training samples, and the rest were used as test

samples. The selected training samples are trained by a random forest algorithm to obtain a malware feature set and a benign software feature set. The test sample is input into the random forest model, compared with the already trained software feature set, and each node is voted to determine whether the software is malicious or benign.

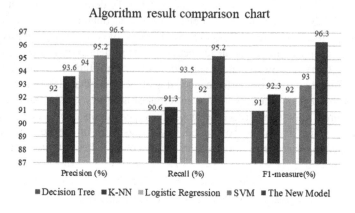

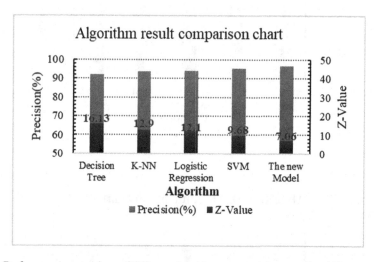

Fig. 6. Performance comparison of different algorithms on precision, recall and F1-score metrics.

Analysis of test results: As shown in Fig. 6, The new model is compared with the detection results of other algorithms. The Fig. 7 shows the performance comparison between the malware and benign software in regards to the performance metrics - accuracy, recall, and F1-measure. Within the Fig. 7, the x axis denotes the different algorithm, the y axis on the left represents the recall, precision, and F1-measure, and the y axis on the right hand side denotes the Z values, which is employed to quantitatively distinguish the performance of different approaches in different percentages of labelled experiment data with regards to precision or recall or F1-score. Mathematically, it is

defined as Z(X) = [expected(X) − X]/standard-deviation[Decision Tree (X), K-NN (X), SVM (X), Logistic Regression (X), The Proposed Model (X)] for three different approaches, where X represents the various performance benchmarks (precision or recall or F1-measure) considering labelled experimental data. In addition, expected (X) = 100 for each case, since this is the ideal situation. To demonstrate the performance comparison clearly, the Z values are labelled in each bar diagram. Through Fig. 6, the accuracy rate, recall rate and F1-score comparison of various algorithms can be obtained: The proposed model achieves better results in detection accuracy than other algorithms in Android software, and is superior to other algorithms in terms of recall rate and f1-measure. In terms of performance, it can be seen that the proposed model algorithm is better than other algorithms in terms of accuracy rate, recall rate and F1-measrue. Figure 7 shows the difference in z-values (performance) obtained for different models, from which it can be concluded that the z-value of the new model is the smallest, indicating that the performance of the detection is better than other models.

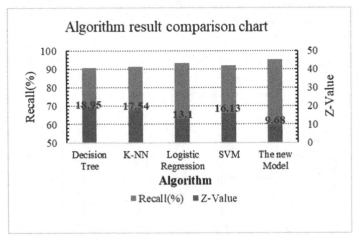

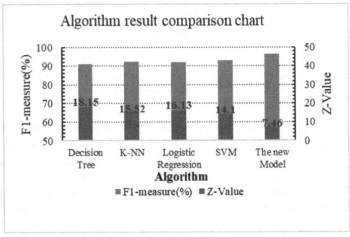

Fig. 7 Performance comparison of z-values of algorithms

6 Conclusion

This paper proposes an ensemble learning based malware detection technology based on QuorumChain framework, which uses the QuorumChain to realize data sharing between nodes. The QuorumChain is a kind of blockchain. On the basis of the distributed data storage of the blockchain, it has its own characteristics of data sharing between nodes. In data transmission, it can only be transmitted between nodes that allow each other, which ensures the difference of data between nodes in the data processing stage, making detection more efficient and accurate.

The experimental results show that the model outperforms other algorithms in malware detection accuracy, recall rate and model accuracy F1-measure, and its performance z-value is better than other algorithms. The preliminary detection results of the model have achieved good results, but in order to further improve the detection efficiency, we can improve the model by improving the ensemble learning module or using more samples and more suitable algorithms.

References

1. IDC.COM: Smartphone market share. https://www.idc.com/promo/smartphone-market-share/os. Accessed 30 June 2019
2. National Computer Network Emergency Technical Processing Coordination Center. Specification for mobile internet malicious code, YD/T 2439 (2012)
3. Chen, K., Wang, P., Lee, Y., Wang, X., Zhang, N., Huang, H., et al.: Finding unknown malice in 10 seconds: mass vetting for new threats at the Google-play scale. In: Usenix Conference on Security Symposium. USENIX Association (2015)
4. Zhang, M., Duan, Y., Yin, H., Zhao, Z.: Semantics-aware Android malware classification using weighted contextual API dependency graphs (2014)
5. Xiao, X., Xiao, X., Jiang, Y., Liu, X., Ye, R.: Identifying Android malware with system call co-occurrence matrices. Trans. Emerg. Telecommun. Technol. **27**(5), 675–684 (2016)
6. Di Cerbo, F., Girardello, A., Michahelles, F., Voronkova, S.: Detection of malicious applications on Android OS. In: Sako, H., Franke, K.Y., Saitoh, S. (eds.) IWCF 2010. LNCS, vol. 6540, pp. 138–149. Springer, Heidelberg (2011). https://doi.org/10.1007/978-3-642-19376-7_12
7. Fuchs, A.P., Chaudhuri, A., Foster, J.S.: SCanDroid: automated security certification of Android applications (2009)
8. Shang, F., Li, Y., Deng, X., He, D.: Android malware detection method based on Naive Bayes and permission correlation algorithm. Cluster Comput. **21**(8), 1–12 (2017)
9. Li, Y., Shen, T., Sun, X., Pan, X., Miao, B.: Detection, classification and characterization of Android malware using API data dependency. In: Thuraisingham, B., Wang, X., Yegneswaran, V. (eds.) SecureComm 2015. LNCS, vol. 164, pp. 23–40. Springer, Cham (2015). https://doi.org/10.1007/978-3-319-28865-9_2
10. Bhatia, T., Kaushal, R.: Malware detection in Android based on dynamic analysis. In: 2017 International Conference on Cyber Security and Protection of Digital Services (Cyber Security). IEEE (2017)
11. Enck, W., Gilbert, P.: TaintDroid: an information flow tracking system for real-time privacy monitoring on smartphones. In: Proceedings of the Usenix Symposium on Operating System

Design and Implementation, OSDI, Vancouver, BC, Canada, 4–6 October 2010, pp. 393–407. DBLP (2010)

12. Bläsing, T., Batyuk, L., Schmidt, A.D., Camtepe, S.A., Albayrak, S.: An Android application sandbox system for suspicious software detection. In: 2010 5th International Conference on Malicious and Unwanted Software (MALWARE). IEEE (2010)

13. Su, M.Y., Fung, K.T., Huang, Y.H., Kang, M.Z., Chung, Y.H.: Detection of Android malware: combined with static analysis and dynamic analysis. In: 2016 International Conference on High Performance Computing & Simulation (HPCS). IEEE (2016)

14. Yuan, Z., Lu, Y., Xue, Y.: Droiddetector: Android malware characterization and detection using deep learning. Tsinghua Sci. Technol. **21**(1), 114–123 (2016)

15. Zhang, X., Hu, D., Fan, Y., Yu, K.: A novel Android malware detection method based on Markov blanket. In: IEEE International Conference on Data Science in Cyberspace. IEEE (2016)

16. Nakamoto, S.: Bitcoin: a peer-to-peer electronic cash system (2008). https://bitcoin.org/bitcoin.pdf 《Consulted》

17. Di Pierro, M.: What is the blockchain? IEEE Comput. Sci. Eng. **19**, 92–95 (2017)

18. Puthal, D., Malik, N.: The blockchain as a decentralized security framework [future directions]. IEEE Consum. Electron. Mag. **7**(2), 18–21 (2018)

19. Thomas, G.D.: Machine learning research: four current directions. AI Mag. **18**(4), 97–136 (1997)

20. Zhihua, Z.: Machine Learning (Chinese Edition), pp. 171–191. Tsinghua University Press, Beijing (2016)

21. Cummins, M., Newman, P.: Probabilistic appearance based navigation and loop closing. In: IEEE International Conference on Robotics & Automation. IEEE (2007)

Asymmetric Information in High-Value Low-Frequency Transactions: Mitigation in Real Estate Using Blockchain

Mark Hoksbergen, Johnny Chan, Gabrielle Peko,
and David Sundaram$^{(\boxtimes)}$

Department of Information Systems and Operations Management,
University of Auckland, Auckland, New Zealand
{m.hoksbergen, jh.chan, g.peko,
d.sundaram}@auckland.ac.nz

Abstract. The saying 'buyer be aware' has been used as an excuse from a seller's perspective to withhold information that could negatively impact a transaction. This asymmetric information is especially prevalent in high–value, low-frequency assets. Using New Zealand real estate as an exemplar to understand the difficulties faced in such a transaction, we delve into the characteristics of the transaction and specifically the asymmetric information that is predominant in the real estate industry. Understanding the processes and stakeholders involved gives us the possibility to introduce blockchain as a system to mitigate asymmetric information. We identify the bottlenecks in the current processes and suggest possible solutions that capitalize on the blockchain characteristics.

Keywords: Transaction · Information asymmetry · Blockchain · New Zealand real estate · Information value chain

1 Introduction

Although trading goods has always been part of human existence, this has invariably created tension between buyers and sellers. This tension is very much present in high-value assets particularly where information is hard to obtain. Asymmetric information has been misused often in history. Plato aptly states that honesty is, for the most part, less profitable than dishonesty. Buyers that infrequently purchase high-value assets through an intermediary are at a particular disadvantage and increased risk of coming to a purchase decision based on unreliable information. Many governments attempt to prevent the misuse of information through legislation. The United States of America has a "Lemon Law" that protects the buyer. In New Zealand (NZ), the Fair Trading Act protects consumers. However opportunistic behaviours flourish and result in financial hardship for the disadvantaged at the fuzzy edges of the margins of the law.

Our research motivation primarily stems from our collective, first-hand experience of the NZ residential housing market and the information asymmetry that exists herein. In addition, the topic is generalizable to many high-value low-frequency transactions

© Springer Nature Switzerland AG 2019
R. Doss et al. (Eds.): FNSS 2019, CCIS 1113, pp. 225–239, 2019.
https://doi.org/10.1007/978-3-030-34353-8_17

(HVLFT) that exhibit the same core problems and issues as a real estate transaction. Our focus is to develop a discussion on asymmetric information and how it can be addressed with the use of blockchain technology. For this to happen, we will first examine the current situation within the NZ real estate industry to understand the characteristics of a typical HVLFT transaction. Then, we clarify asymmetric information in the context of a real estate transaction. Finally, we explore the possibility of utilizing blockchain technology to mitigate and counterbalance asymmetric information that exists between the buyer and seller of a high-value asset.

2 New Zealand Real Estate

The purchase and sale of real estate in NZ is becoming increasingly more complex. A bevy of new rules and regulations has recently been introduced, resulting in more responsibilities being delegated to real estate agents and buyers. For example, new regulations introduced by the Overseas Investment Office restrict overseas buyers purchasing real estate in NZ and there is the 2019 anti-money laundering bill. Also, for buyers there are problems identifying all the relevant documents needed to make an informed decision on a real estate transaction and to further conduct appropriate due diligence. Moreover, transferring titles requires a lawyer to handle any mortgages, wills, title transfer and trust account for the payment.

When purchasing real estate, acquiring accurate and reliable information is a challenge whether one be a real estate agent, developer, seller or buyer. Based on our experience, two core problems exists that relate to the distribution of information pertaining to a property. The first problem is regarding information silos where property information is distributed over several organizations such as government, specialists, councils, and real estate organizations, and there is little motivation to share it. This information is treated as a commodity by these organizations, which makes it costly to obtain and verify for external parties. For example, discrepancies in information can exist between the property file held by Land Information NZ and the Council. Each of these governing organizations sell this information to interested third parties independently and without recourse so there is no additional gain for the organisations to change the status quo. Another example is when a buyer hires specialists during the due diligence period. If the process does not result in an unconditional offer, the property is often relisted for sale and all the prior information collected is lost. In a small market such as NZ, this quite often results in the same specialist producing the same information numerous times on the same property or multiple specialists hired for the same task but providing different outcomes.

The second problem is a verifiability problem, which stems from the various building experts, engineers, surveyors and other experts collecting enormous amounts of information relating to a particular property without the obligation to verify its quality [1]. Thus, creating a lack of accountability for the information provided, which comes with a disclaimer that the organization providing it is not liable for any errors. Examples of such disclaimers can be seen in Table 1.

The verifiability problem is exacerbated when a real estate agent conveys biased, anecdotal information pertaining to a property that is then used as the soft criteria for

purchasing it. The principal intermediary in the NZ context is the real estate agent. Quite often this intermediary does not always have the best interest in mind for both the buyer and the seller for a sale to proceed [2]. The primary interest of the real estate agent is the actual sale of a property. Their key performance indicators (KPIs) are the sales volume, number of potential buyers, and number of listings. These KPIs often are not in alignment with the best interests of the buyer. Once the initial price negotiations are completed, there is usually a deadline to finalise and complete the due diligence by the buyer, increasing the pressure to acquire the relevant information.

Table 1. Disclaimers per report type.

Report type	Disclaimer
LIM report	It remains the responsibility of the seller and buyer to determine the final rate figures of the settlement date and ensure that this is paid in accordance with the requirements of local government's act 2002
	Where information has been supplied to counsel by a third party, it cannot guarantee the accuracy of the information, and it is provided on the understanding that no liability shall arise or be accepted by the cultural for any errors containing therein
Building report	This is a report of visual only, non-invasive inspection of the areas of the building. The inspection did not assess compliance with the NZ building code: including the code's weather tightness requirements for structural aspects

Delving deeper into the practicalities of the due diligence process, the Council holds a property file that contains all the consent notices, forms and code of compliances for a particular property. This information is summarized, along with the legal description, lands use, zoning, utilities, etc., in the Land Information Memorandum (LIM) report, which can be purchased from the Council for approximately NZ$200 to $400. Often, the details in the property file and the LIM are not correlated and have critical information missing. Again, the information presented in a LIM report comes with a disclaimer that the Council is not liable and cannot and will not guarantee the accuracy of the information. Furthermore, the LIM report does not have any page numbers which means pages can be inserted or removed without anybody being the wiser and thus making it susceptible to manipulation. Due diligence also requires the buyer to obtain the land title that gives caveats, covenants and more lodged with the title. Then there are a multitude of other reports that can be requested, such as builders report, valuation report, drainage report, moisture report, electrician report, asbestos report, soil test, archaeological test, and geo-tech report. These reports are commissioned by the buyer and not reported to the Council often resulting in the same report being requested multiple times or overlooked completely.

The real estate industry typically has many intermediaries before a satisfactory transaction between a seller and a buyer can be concluded. Moreover, the buyer, is negatively impacted by the siloed information and verifiability problems. This adds to the complexity of obtaining the necessary information needed to conduct proper due diligence in a short time frame and make an informed decision.

In addition to the above mentioned challenges in gathering verifiable information, often the most valuable information is the information that is not shared between the seller and buyer. This information is known by the seller and not shared so the sale price is not compromised. When buying a property, there are many indicators that could assist the sale, which are often difficult to obtain. These indicators are often verbally transmitted, unrecorded and eventually lost. Private information known only to the seller could often flesh out the details of the property being sold. This information is often very diverse and could range from information about a new hot-water cylinder installed, a new alarm system, a recent repaint of the house, and the type of solar system in place. This information would have a date, name of installer, price, and warranty details. Capturing such information begets the possibility of adding "owners" data to the property file that could then be linked to the LIM.

In order to understand the characteristics of a real estate transaction, we first posit that it can be defined as a HVLFT and it is generalizable to other similar transactions. In the following sections, we define a HVLFT and the effects on the price equilibrium. This helps to clarify our understanding of the information and knowledge that is essential to make an informed decision and identify the stakeholders. Next, we delve into the role that asymmetric information has on a transaction and the stakeholders involved. We finally suggest blockchain as a possible solution in limiting the risk and cost to a potential buyer when engaging in a HVLFT.

3   High-Value Low-Frequency Transactions

There are a number of disciplines that view a transaction in a different light. Yet, there is only a modest reference to transactions in the literature, most of which is based on transaction cost economics [3]. Transaction cost is a cost in making any economic trade when participating in a market. These costs in our context, consist of three main types; search and information cost, bargaining cost, and enforcement cost. In the context of business, at the most basic level, a transaction can be defined as the exchange of goods or services between a buyer and a seller. Transactions can be further defined by including the transfer of goods, services, and money, the transfer of title and/or possession and, the transfer of exchange rights [4].

For our purpose, we include two additional characteristics to the transaction; *high-value* and *low-frequency*. To consider a transaction to be of *high-value*, the NZ Ministry of Justice defines a high-value asset as any asset that exceeds the value of NZD $10,000 [5]. However, it can be argued that this definition is unequivocal and gives rise to many exceptions. Alternatively, we can view a high-value asset from the buyer's and seller's perspective and what they perceive as a high-value transaction. To provide an example, Bill Gates might perceive that the purchase of a new car is a relatively low-value transaction whilst a typical student will view the purchase as an extremely high-value transaction. So we will use the personal perception of a transaction as our

determinant. In terms of *low-frequency,* this is yet another disputable term. Again, we adopt the view of low-frequency as perceived between the two parties. As a guide, we suggest assets that are purchased less than ten times in a lifetime as low-frequency, which may not apply to professional traders, agents, and specialists.

Summarizing the aforementioned into a single definition, we define a HVLFT as:

> *A transaction where the buyer and seller perceive the asset to be of high-value, and the number of transactions are only a few over a person's lifetime.*

3.1 HVLFT's Costs, Characteristics, and Emotions

In a HVLFT, we are committing a large part of our financial wealth on an asset that we might only buy or sell a few times in our life. To understand the differences between a regular transaction and a HVLFT, we need to highlight these differences from the buyer's and seller's perspectives. For this, we have chosen a multi-lense approach. We attempt to unravel the main costs, characteristics, and emotions affecting the buyer and seller. This is by no means a complete overview but it highlights the main differences. Using transaction costs [3] to segment a transaction into three main costs, we use this as a lens to examine the differences between a regular transaction and a HVLFT. There are certain parameters that change when a transaction is scaled up by their two core variables (high-value and low-frequency). Also, we use two other variables, which are harder to define, trust and willingness to accept risk. These variables are classified as emotional responses [6]. For trust and the willingness to accept risk, within the context of a transaction between two unrelated stakeholders, we adhere to the viewpoint [7] that an individual will maximize their own utility first above that of a group or organization. A high-value transaction increases the financial risk a buyer exposed to and also decreases the trust from the seller's perspective. In layman terms, the "stakes have increased." We often find that in a high-value transaction the number of stakeholders involved also increases. This is mainly on the buyer's side due to specialists needed to verify the information. Consequently, the cost of accurate information and legal enforcement costs increase due to the greater complexity of the transaction (Fig. 1). In summary, if we accept the premise that information and knowledge affect the above mentioned emotions, consequences, and costs, we realize the gravity of asymmetric information. That, ultimately affects the price and the level of exposure to risk.

3.2 Price Equilibrium in a Transaction

One of the main factors that affects price elasticity in low-frequency traded goods is the number of substitutes available, which in turn impacts the price elasticity of a transaction [8]. This is linked to the comparability of the goods and services and affects the buyer as they try to ascertain the market price while the seller is liable for the bargaining cost in the form of the agent's commission.

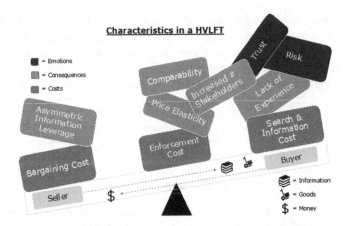

Fig. 1. Characteristics in a HVLFT between a seller and buyer

Ultimately from a seller's and buyer's point of view, a transaction can be represented by three lines on a graph. The lowest line representing an intrinsic minimum the seller is willing to accept, and the highest line is the asking price or the maximum a buyer is willing to pay. The middle line represents the actual market and would be the best scenario for the buyer and the seller (win/win) as illustrated in Fig. 2. When either the seller or the buyer deviate from the market they are, potentially, left with a financial liability, and moreover, it is the buyer who is exposed to this risk.

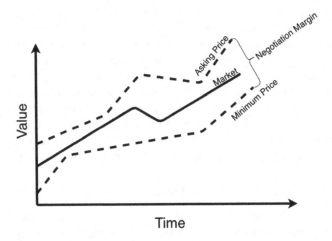

Fig. 2. Graphic representation of the price tension in a negotiation

To achieve a purchase price that is closer to the actual market price, information is crucial. One significant drawback in a HVLFT is the asymmetric information, which (usually) favours the seller, and the inexperience that comes with the low-frequency of

the transaction and the associated risk. Also, ascertaining the average price can be challenging due to the lack of comparable assets. Still, the seller would have the advantage because they have a more profound knowledge of the asset.

There is a correlation between the risk and the price a buyer is willing to pay. A buyer will counteract the risk by building in a margin to offset the perceived risk [9]. From the seller's perspective, the potential in lost margin could be a motivation to share more information with the buyer.

4 Asymmetric Information

Economists have discussed problems relating to information asymmetry since the eighteenth century. Information is often not perfect and obtaining information is costly. There is always a level of information asymmetry in competitive markets, the extent of which is affected by actions of firms and individuals. Asymmetric information gives advantages to one party while disadvantaging the other.

To understand the context of asymmetric information within a HVLFT, we need to understand the difficulties buyers face when purchasing a HVLF asset. One assumption is that due to the low-frequency purchasing of an asset, the possibility of single, double, or triple loop learning is less applicable [10]. Because the opportunities are few to reflect on mistakes, question whether we are doing things in the right way and adapt our actions, assumptions, and context are few.

Given, that a HVLFT is a result, the chance of repeating a similar transaction is unlikely, thus significantly limiting the opportunity to learn from experience. Due to the infrequency, buyers in particular have less chance of learning from their mistakes. This means that information given to a buyer needs to adhere to several criteria to address its asymmetry. According to Maltz [11], these criteria are; codability, comprehensibility, reliability, relevancy, and volume of information. In other words, the data needs to be trusted, verifiable, understandable, and traceable. It needs to be able to stand up in a court of law. In gathering this information, it also needs to be affordable and able to be collected within the negotiation time constraint.

In the context of a transaction, sellers often have the advantage of information asymmetry over the buyer [12]. This can lead to sub-optimal decision-making by the buyer. Asymmetric information in transactions has been studied in several industries such as the automotive industry [13, 14], business process design [15] and residential housing markets [16, 17]. They find information asymmetry usually benefits the seller and disadvantages the buyer.

The disclosure of asymmetric information also negatively impacts the implicit price of an asset. By contrast, it has been argued that to counterbalance perceived information asymmetry, a buyer adds a risk component into the transaction to cover potential information oversights [18].

Purchasing a high-value asset frequently leads to a more significant disadvantage for the buyer. This is especially applicable to HVLFT. The information asymmetry between a buyer and seller of a real estate transaction is shown in Fig. 3. The model addresses the tension between tacit and explicit information, the input of verifying data

through specialists, and the role of government in the transaction. If a conflict occurs in the transaction, the noise created in the channel can be hard to resolve.

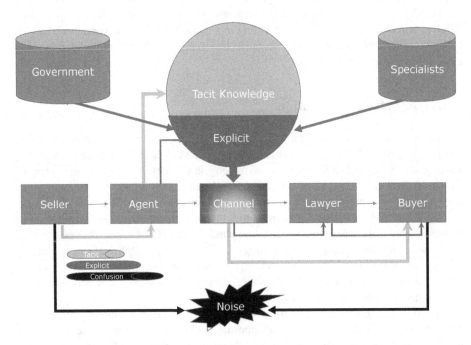

Fig. 3. Adapted model of communication from Shanon and Weaver [19]

Applying the model of communication (Fig. 3) to the NZ real estate industry, the sender represents the seller, the encoder the agent, the decoder the lawyer and the receiver the buyer. High-value assets are often traded through an intermediary and explicit information is checked by lawyers. There is a disincentive for the seller or agent to codify information if it is not beneficial to the transaction from their perspective. The noise in the channel is increased by the convoluted information, which is hard to verify and significantly impacts the decision-making process of the buyer. The challenge faced by buyers is that explicit information is often convoluted and derived from multiple sources. Furthermore, tacit information is often verbally transmitted by the seller or agent., so if a conflict occurs in the transaction, the noise created in the channel can be hard to legally resolve.

[20] argues that the main goal of knowledge is to progress towards understanding and then to wisdom. [21] developed a model that illustrates the process of formalizing facts and ideas and processing them into information, which is converted to knowledge that ultimately leads to wisdom. In addition, [20] posits that discovery starts with observations that can be converted to data that leads to information. Once knowledge is gained, it needs to be understood before it develops into wisdom. [22] suggests that contextualization is the primary driving force to fathom the transformation from data to knowledge. Along the same lines, [23] elaborates on [20] model adding the

environment as a prerequisite to data and vision as the final result of wisdom. [24] describes a path that builds knowledge from understanding data and information in the past, creating an experience to future knowledge and wisdom thereby developing a model based on four building blocks: data, information, knowledge, and wisdom. Although the concept of four steps from data to wisdom is accepted, the authors posit that a circular model may be more appropriate. As soon as information changes, knowledge can change. This in turn requires us to revisit the data and reassess our wisdom, and so on.

Granted that the models mentioned above enable understanding, it can be argued that they lack a degree of practicality. [25] argues that theory practicality is associated with measurability; this highlights a core problem in gaining the necessary knowledge and reducing information asymmetry.

The hedonic pricing method can also be used as an insight into the use of asymmetric information in a transaction. Hedonic pricing is a model which identifies price factors according to the premise that price is determined by both the internal characteristics of the good being sold and the external factors affecting it [16]. There have been several studies that have used hedonic pricing to correlate environment factors to house prices. [17] studied the effect of the disclosure of airport noise and the implicit price of a house. Concluding that house prices were negatively affected when noise impacts from the airport were disclosed to potential buyers. This affects the motivation for sellers to disclose environmental issues. The assumption made by [26] that the availability and quality of information available for the decision-making process significantly impacts the final price of a transaction. From the perspective of the seller, asymmetric information will prevail over symmetric information.

5 Stakeholders

We have identified six main stakeholders. Naturally, within any transaction, there is a buyer and a seller. In a HVLFT, there are multiple stakeholders including legal representatives for the buyer and the seller to actualize the transaction and lower the risk. Often the buyer will employ several specialists to verify and check the information received. The seller will often work through an intermediary or agent. Finally, the government is involved, it is responsible for rules relating to the transaction and sometimes can impact the market significantly. In Fig. 4 the different stakeholders are represented and how price and information are influenced by their interaction in a HVLFT. We have combined the seller and the agent together to present a clearer overview.

6 Theoretical Framework

The theoretical framework illustrated in Fig. 5 is a synthesis of the DIKW Model [20] and Skill Acquisition model [27], suggesting that if we can educate the novice buyer into asking expert questions the risks can be reduced. Giving them a roadmap to avoid potential errors caused by inexperience further reduces the risk. The model (Fig. 5)

represents the five steps of acquiring skills from novice to expert as well as showing the development of data, information, knowledge, and wisdom (DIKW). The model needs to be seen in the context of a single transaction. If trust in the presented information is high, the risk in the transaction will decrease. As mentioned, in the majority of HVLFT the buyer is a novice and increasing the user-friendliness on how the DIKW process is presented, will address the understanding of the asset and consequently reduce the risk in the transaction.

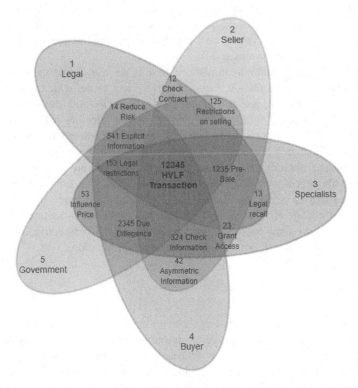

Fig. 4. Venn diagram of the interaction between stakeholders in a HVLFT

7 Knowledge Management Utilizing Blockchain

As discussed, a HVLFT has different characteristics to regular transactions including information asymmetry. This information asymmetry in HVLFT causes a number of problems. First, our primary practical problem is managing information in the context of a transaction. Knowledge-supporting information systems are often linked to knowledge management in the context of an organization and not an industry [25, 28–31]. This also affects the motivation to fill the database with relevant information. Within a market, information and knowledge are seen as a competitive advantage [32]. An organization motivated by this competitive advantage can be motivated to invest in knowledge management systems.

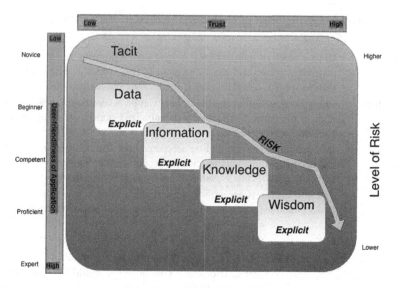

Fig. 5. Risk reduction model for HVLF transactions

Second, the problem within a HVLFT is asymmetric information on a specific asset and the lack of motivation of an entire industry to change these dynamics. It is not beneficial to all parties involved in the transaction to disclose all relevant information. Within the margins of the negotiation, information can directly affect the outcome and price [33].

And third, in a HVLFT, the buyer of the asset is more often the entity that will use the knowledge management system the least and benefit the most. Also, within a HVLFT, the low-frequency deters investment in a knowledge management system from a buyer's perspective. This low-frequency aspect also deters most buyers to invest in gaining the required knowledge to minimize risk. At this point, we need to note that a buyer is not a trader, and we assume that buyers for a specific asset class only transact occasionally and will be the end-user of the asset [34].

Blockchain, a technology that has grown in popularity with Bitcoin since 2008 [35], is enjoying a blossoming beyond cryptocurrency and the transfer of money. It can be used as an architecture to support many types of transactions, from logging an event to signing a document, to voting, to allocating energy between parties, and far beyond. The annual blockchain enterprise application market is set to reach 19.9 billion by 2025 [36].

Bitcoin development is often referred to as Blockchain 1.0 for digital currencies, Blockchain 2.0 for digital finance, and Blockchain 3.0 for a digital society. [37]. Blockchain 1.0 developed in 2008 and 2.0 and 3.0 emerging almost parallel in 2015 [38]. This emergent technology is based on the concept of a distributed, immutable, historical, and verifiable ledger.

Using blockchain as a distributed ledger seems to fulfil most of the characteristics needed to reduce the cost and increase the reliability of information needed in a HVLFT. According to [39] there are four core characteristics as discussed in Table 2.

Table 2. Characteristics of blockchain in real estate adapted from [39]

Characteristics	Summary
Immutable	Once a block of information is logged onto a property this becomes a permanent record and cannot be altered. This creates an historical record that provides a complete picture (permanent and tamper-proof)
Decentralized	Information stored on the blockchain can be accessed and copied by any node in the network. Creating a decentralized storage providing multiple backups (networked copies)
Consensus driven	Information can be verified by appointed nodes, thus increasing reliability and accountability. This mechanism works without the use of a central authority or an explicit trust-granting agent (trust verification)
Transparent	The information is accessible and cost effective and creates provenance under which asset lifetimes can be tracked (full transaction history)

Given the characteristics of HVLVT, blockchain could be a suitable technology to address a number of our concerns. Blockchain technology could be used to store verifiable information onto permissioned blockchains. This would address information asymmetry in a variety of ways. First, all information would be time-stamped, situating the data in a historical context. Second, the information would be verifiable since it will only be entered through trusted nodes. Third, it can give a legal standing to the buyer in case the information used to make a purchase decision was incorrect. And, fourth, information that had previously been collected, does not have to be sourced again, thus saving time and money.

The problem that arises is the motivational aspect of the stakeholders to participate in the information exchange on the blockchain. We need to understand their financial loss or gain in sharing information and list motivations per stakeholder to share verifiable information and how we can verify this information (Table 3).

Table 3. Stakeholder resistance

Stakeholders	Resistance
Seller	Increased transparency could lead to a lower price
Agent	Increased transparency and verifiability could lead to lower margins and lower market share. They become less relevant in the transaction
Specialist	Less opportunity to 'resell' information
Government	Increased insights into the development and detailed information per household
Legal	Lower transaction costs and reduced legal work from transactions that have gone wrong
Buyer	Lower cost in gaining verifiable and trustworthy information. Reduced risk in the transaction

Initially, from the six stakeholders, only two benefit directly from the introduction of blockchain. The government increases its detailed information per household. The

buyer lowers his cost in the due diligence process and benefits from the increased amount of verifiable data.

The government would need to be the catalyst to drive change, motivating the buyer to change their behaviour. As mentioned, lowering risk and initial cost in a transaction is the driving force to initiate change in this space. Counterbalancing the asymmetric information and increasing the knowledge and trust on the asset that will be transacted, would be the largest contributors to reducing that risk.

As an incentive for the seller to participate in the transparency of the asset, the reduction of risk could have a positive impact on the sales price. There is a correlation between risk and the price a buyer is willing to offer. A buyer will counteract the risk by building in a margin to offset the perceived risk [9]. There is also a motivation for the specialists involved if we can convince sellers prior to listing their property to acquire and register the reports on the blockchain in order to present an in-depth picture of their asset, thus reducing the buyer's risk. If we can motivate the buyer and seller to utilize blockchain for information sharing and use government as an initiator, we can counterbalance information asymmetry within HVLFT.

8 Conclusion and Discussion

This paper focusses on the buyer in the context of a HVLFT. We have recognized four major hurdles with a transaction that negatively impact the buyer and heightens their risk. This is especially applicable in a real estate transaction. First, the lack of experience of the buyer due to the characteristics of a HVLFT. The novice buyer cannot rely on a learning curve and does not possess the expertise necessary to negate the risks posed by such a transaction. Second, there is no immediate incentive from a seller point of view to counterbalance asymmetric information. There are a raft of specialists that the buyer can employ to counterbalance the information but there is a direct initial cost. In addition, there is a cost to check the information, given the lack of accountability on behalf of the specialists and government. Within this process, there is also a lot of information being lost if a buyer does not complete the transaction. This information resides with the buyer, it is not stored or made public and the same information can be sold again to the next potential buyer. The third hurdle is ascertaining the actual market price due to the lack of comparability of the asset. And finally, a HVLFT is more susceptible to government changes in regulations. A few recent practical examples in NZ are the change in anti-money laundering laws, the new regulation restricting foreign buyers of purchasing residential property, and the mandatory earthquake strengthening.

We need to create an information structure and model for all the relevant information involved in HVLFT decision-making. We need to create a decision model for a buyer based on a permissioned blockchain ledger, as a way to efficiently and securely organize explicit information. Furthermore, deploying a personalized database for property owners to the current information available allows for the transfer of tacit information to verifiable explicit information direct from the seller to the buyer.

Hopefully, this research will lead to a practical working platform for councils, real estate agencies, buyers and sellers of real estate and drastically cut the cost of paperwork involved in the due diligence of buying real estate in NZ and increase the

accuracy and security of information presented to the buyer. This could also ensure that information on a property is not lost and repeated multiple times and that there is a chain of information that is traceable back to the origins of the property. The personalized database for property owners would allow owners to add detailed information to the current ledger and possibly improve the code of practice for installers and improve warranty claims.

References

1. Lillrank, P.: The quality of information. Int. J. Qual. Reliab. Manag. **20**, 691–703 (2003)
2. Allmon, D.E., Grant, J.: Real estate sales agents and the code of ethics: a voice stress analysis. J. Bus. Ethics **9**, 807–812 (1990). https://doi.org/10.1007/BF00383279
3. Williamson, O.E.: Chapter 3 transaction cost economics (1989). https://linkinghub.elsevier.com/retrieve/pii/S1573448X8901006X
4. BusinessDictionary.com: What is a market? Definition and meaning. http://www.businessdictionary.com/definition/transaction.html
5. NZ Government Justice: Businesses trading in high value goods and AML/CFT. New Zealand Ministry of Justice. https://www.justice.govt.nz/justice-sector-policy/key-initiatives/aml-cft/info-for-businesses/high-value-goods/#why
6. Frederiks, E.R., Stenner, K., Hobman, E.V.: Household energy use: applying behavioural economics to understand consumer decision-making and behaviour. Renew. Sustain. Energy Rev. **41**, 1385–1394 (2015). https://doi.org/10.1016/J.RSER.2014.09.026
7. North, D.C.: The new institutional economics (1986)
8. Ellison, G., Ellison, S.F.: Search, Obfuscation, and Price Elasticities on the Internet. Econometrica. **77**, 427–452 (2004). https://doi.org/10.2139/ssrn.564742
9. Kim, D.J., Ferrin, D.L., Rao, H.R.: A trust-based consumer decision-making model in electronic commerce: the role of trust, perceived risk, and their antecedents. Decis. Support Syst. **44**, 544–564 (2008). https://doi.org/10.1016/J.DSS.2007.07.001
10. Argyris, C., Schon, D.A.: Theory in Practice: Increasing Professional Effectiveness. Jossey-Bass, Oxford (1974)
11. Maltz, E.: Is all communication created equal? An investigation into the effects of communication mode on perceived information quality. J. Prod. Innov. Manag. **17**, 110–127 (2000). https://doi.org/10.1111/1540-5885.1720110
12. Neelawala, S.: Asymmetric information between buyers and sellers in the residential property market: a hedonic property valuation approach (2014)
13. Hendel, I., Lizzeri, A.: Adverse selection in durable goods markets. American Economic Association (1999)
14. Akerlof, G.A.: The market for "Lemons": quality uncertainty and the market mechanism. Q. J. Econ. **84**, 488–500 (1970). Oxford University Press Stable
15. Seidmann, A., Sundararajan, A.: The effects of task and information asymmetry on business process redesign. Int. J. Prod. Econ. **50**, 117–128 (1997). https://doi.org/10.1016/s0925-5273(97)00037-6
16. Rosen, S.: Hedonic prices and implicit markets: product differentiation in pure competition. J. Polit. Econ. **82**(1), 34–55 (1974)
17. Pope, J.C.: Buyer information and the hedonic: the impact of a seller disclosure on the implicit price for airport noise. J. Urban Econ. **63**, 498–516 (2008). https://doi.org/10.1016/J.JUE.2007.03.003

18. Easley, D., Hvidkjaer, S., O'hara, M.: Is information risk a determinant of asset returns? American Finance Association (2002)
19. Shanon, C.E., Weaver, W.: The Mathematical Theory of Communication. University of Illinois Press, Champaign (1949)
20. Ackoff, R.L.: From data to wisdom. J. Appl. Syst. Anal. **16**, 3–9 (1989). citeulike-article-id:6930744
21. Cleveland, H.: Information as a resource. Futurist. **16**, 34–39 (1982)
22. Tuomi, I.: Data is more than knowledge: implications of the reversed knowledge hierarchy for knowledge. Manag. Inf. Syst. **16**, 107–121 (1999). https://doi.org/10.1109/HICSS.1999.772795
23. Carpenter, S., Cannady, J.: Tool for sharing and assessing models of fusion-based space transportation systems. In: 40th AIAA/ASME/SAE/ASEE Joint Propulsion Conference and Exhibit. American Institute of Aeronautics and Astronautics, Reston (2004)
24. Sarah Fraser: News, views and challenges from Sarah Fraser: model 4: data, information, knowledge, wisdom, http://spreadgoodpractice.blogspot.com/2010/12/model-4-data-information-knowledge.html
25. Harlow, H.: The effect of tacit knowledge on firm performance. J. Knowl. Manag. **12**, 148–163 (2008). https://doi.org/10.1108/13673270810852458
26. Jullien, B., Mariotti, T.: Auction and the informed seller problem. Games Econ. Behav. **56**, 225–258 (2006). https://doi.org/10.1016/j.geb.2005.08.008
27. Dreyfus, H.L., Dreyfus, S.E., Koschmann, T.D.: Book Review Five Stages in the Acquisition of Expertise. Basil Blackweli, Oxford (1986)
28. Spender, J.-C.: Making knowledge the basis of a dynamic theory of the firm. Strat. Manag. J. **17**, 45–62 (1996)
29. Lakshman, C.: Organizational knowledge leadership: a grounded theory approach. Leadersh. Organ. Dev. J. **28**, 51–75 (2007). https://doi.org/10.1108/01437730710718245
30. Takeuchi, H., Nonaka, I.: Theory of organizational knowledge creation. In: Morey, D., Maybury, M., Thuraisingham, B. (eds.) Knowledge Management: Classic and Contemporary Works, pp. 139–182. MIT Press, Cambridge (2002)
31. Goh, S.C.: Managing effective knowledge transfer: an integrative framework and some practice implications. J. Knowl. Manag. **6**, 23–30 (2002). https://doi.org/10.1108/13673270210417664
32. Teece, D.J.: Capturing value from knowledge assets: the new economy, markets for know-how, and intangible assets. Calif. Manag. Rev. **40**, 55–79 (2012). https://doi.org/10.2307/41165943
33. Chua, A.: Knowledge sharing: a game people play. Aslib Proc. **55**, 117–129 (2003). https://doi.org/10.1108/00012530310472615
34. Bamberg, G., et al.: Agency Theory, Information, and Incentives. Springer, Heidelberg (2011). https://doi.org/10.1007/978-3-642-75060-1
35. Nakamoto, S.: Bitcoin: a peer-to-peer electronic cash system. Satoshi Nakamoto Institute. www.cryptovest.co.uk
36. Hansen, M.D., Kokal, M.: The coming blockchain disruption: trust without the "middleman" (2017)
37. Swan, M.: Blockchain: Blueprint for A New Economy. O'Reilly Media Inc., Sebastopol (2015)
38. Zhao, J.L., Fan, S., Yan, J.: Overview of business innovations and research opportunities in blockchain and introduction to the special issue. Financ. Innov. **2**, 28 (2016). https://doi.org/10.1186/s40854-016-0049-2
39. Sultan, K., Ruhi, U., Lakhani, R.: Conceptualizing blockchains: characteristics & applications. In: 11th IADIS International Conference Information Systems, p. 57 (2018)

Critical Evaluation of RFID Applications in Healthcare

Yu-Ju Tu[1], Huihui Chi[2], Wei Zhou[2], Gaurav Kapoor[3], Enes Eryarsoy[4], and Selwyn Piramuthu[5(✉)]

[1] Management Information Systems, National Chengchi University, Taipei, Taiwan
[2] Information and Operations Management, ESCP Europe, Paris, France
[3] Infocomm Technology, Singapore Institute of Technology, Singapore, Singapore
[4] Information Systems, Istanbul Sehir University, Istanbul, Turkey
[5] ISOM, University of Florida, Gainesville, FL, USA
selwyn@ufl.edu

Abstract. The past two decades have witnessed the explosive growth of RFID-based applications in healthcare settings. This includes applications that are related to ePedigree for pharmaceuticals, asset tracking and workflow management in hospitals, pervasive healthcare, among others. We consider some of the benefits that are provided by RFID-based systems in healthcare settings. We then provide a critical evaluation of some of the challenges that are faced by these applications.

1 Introduction and RFID-Based Systems in Healthcare

RFID (Radio Frequency IDentification) tags belong to a class of auto-identification technologies that also include barcode (e.g., Zhou 2009). These resource-constrained tags have minimal processing power as well as some data storage capability. Of the three main types of tags that include active tags, semi-passive tags, and passive tags, the latter is the most popular primarily because of cost and form factor reasons. Some of the characteristics of RFID tags include the ability to store and process data, being able to communicate with a reader that may not be in direct line-of-sight, durability in harsh environments, batch readability, storage of item-level information, ability to carry on a two-way conversation with a reader, among others. The unique identifier in each RFID tag allows for uniquely identifying tagged items and the ability to communicate with a reader allows for process automation.

Item-level information has been beneficially used in ePedigree systems in the pharmaceutical industry to track and trace individual items (e.g., Coustasse et al. 2010). Item-level information has also been used in hospitals to manage process flows (e.g., Gonzalez et al. 2006b), manage inventory (e.g., Meiller et al. 2011), provide automated data for electronic medical record systems (e.g., Mongan et al. 2016), facilitate patient management systems (e.g., Chowdhury and Khosla 2007), as well as other applications such as in surgical sponges (e.g., Williams et al. 2014). RFID is one of the core technologies in pervasive healthcare applications (e.g., Thuemmler et al. 2009).

© Springer Nature Switzerland AG 2019
R. Doss et al. (Eds.): FNSS 2019, CCIS 1113, pp. 240–248, 2019.
https://doi.org/10.1007/978-3-030-34353-8_18

Given the flexibility provided by RFID-based systems, its healthcare applications span a large functional range that include automated data collection, identification, sensing, tracking, as well as managing assets and entities such as blood samples, drugs, hospital equipment, patients, among others. RFID-based systems have also been used to address effectiveness of care, operational efficiencies, patient safety, and quality of care.

At hospitals, RFID-based systems are used to link mother with her newborn by storing the mother's information in the baby's tag. This allows for quick identification and matching of the mother with her baby. Moreover, RFID-based devices such as the monitors that are used around the ankle or wrist of babies help ensure the safety of babies in hospitals. For example, when a baby with this tag is moved near a hospital door, the door is programmed to automatically close. This prevents kidnapping incidents in pediatric and maternity rooms.

RFID tags that are on patients and hospital staff (e.g., on their identification badges) facilitate automatic provision of data for electronic medical records systems. Similarly, RFID tagged medical equipment and supplies are readily tracked and traced. This results in increased efficiency, thereby lowering the costs and increasing the offered service quality. From a survey of published literature, Coustasse and Tomblin (2013) found that RFID use in healthcare settings lead to several benefits and improvements such as the ability to monitor the ambient conditions of perishable and heat-sensitive items, less shrinkage due to theft or misplacement, improved staff productivity and enhanced quality improvement, labor savings and error reduction such as through the use of RFID-enabled surgical sponge (e.g., Williams et al. 2014), better knowledge on equipment availability, improved business processes and workflow, ability to track and manage mobile assets as well as high-cost devices and supplies in real-time, tracking of blood samples, and ensuring compatibility of blood transfusions through RFID tags on the sample as well as the patient.

RFID-based systems have therefore been effectively used in healthcare settings to reduce costs through improved process efficiency and reduced equipment shrinkage that includes misplacement and theft, improve patient safety, simplified patient billing due to linkage with electronic medical record system, and improve supply chain effectiveness. RFID also helps with the automation of processes, thereby removal of human factors from clinician workflow, better inventory management and reduction in stock out situations. Clearly, there are numerous advantages to using RFID-based systems in healthcare settings. On the other hand, these advantages don't come without related issues. We discuss some of these in the next section.

2   Challenges and Issues with RFID Use in Healthcare

Although there has been a steady increase in the number of RFID-based systems that have been implemented in healthcare settings over the last decade, barriers to widespread and rapid adoption include the additional cost of such implementations, unclear return on investment, and competition from other strate-

gic initiatives. Cost has been used as an excuse for resistance to adopt RFID-based systems. When invaluable human lives are at stake, the marginal cost of implementing RFID-based systems pales in comparison to the indispensable and expensive equipments that are used in healthcare settings. Moreover, such arguments only look at one side of the coin - the cost side. Rarely do such line-of-thought include the other side of the coin - the benefit side. When both costs and benefits are simultaneously considered, it has been shown (e.g., Piramuthu et al. 2014) that RFID-based systems win out in a large number of cases because of the tremendous amount of benefits that accrue due to the real-time item-level visibility of tagged items.

The unclear return on investment is due to the fact that we are at the early stage of RFID adoption. This concern will vanish as more and more healthcare institutions start implementing RFID-based systems. As for competition with other strategic initiatives, it is a matter of priority since real-time visibility is of paramount importance in healthcare settings.

There are other issues that are not specific to the healthcare context. For a detailed explanation of these issues, the reader is referred to Kapoor et al. (2009). Specifically, these concerns include ownership transfer issues, privacy/security issues, back-end system bottleneck issues, risk of obsolescence, read rate error, economic disincentives to share item-level information across institutions, and evolving standards. We update this list with a few more concerns that we list and discuss below.

2.1   Data Volume

Each RFID tag generates a large amount of data since it is in real-time contact with reader(s) and each communication instance generates data. Therefore, generation of large volumes of data in RFID-based systems is not difficult. Such data need to be efficiently stored somewhere (e.g., Fazzinga et al. 2009). It is a useless exercise to efficiently store data, when such data are not processed for some useful purpose. Analysis of such massive amounts of RFID-generated data sets is a significant issue in RFID-based systems (e.g., Gonzalez et al. 2006a).

With advanced sensing technologies, RFID is capable of capturing data in a wide range of healthcare-related activities (e.g., Alvarez Lopez et al. 2018). RFID enables autonomous data collection during patient location tracking, patient identification, patient medication and monitoring, as well as drug and medical equipment inventory management processes. For example, RFID in sensor-enabled pill allows for the provision of personalized treatment to individual patients (Mathew et al. 2018). On consumption, such a pill extracts vital organ conditions of the patient and transmits the information to the information system for analysis. It is understandable that patient health is reliant on taking medications regularly. To this end, RFID sensor-based pill bottles (Mathew et al. 2018) are used to ensure that patients take the right dose of medication at the right time. Similarly, sensor-based RFID tags can be used to automatically record vital patient information such as heart rate and blood pressure along with the patient's location and time. These signify the generation of large

volumes of related data that need to be stored and processed for actionable intelligence. Moreover, the extraction of accurate and relevant healthcare information takes time. And, timeliness is generally considered to be very important with respect to healthcare decision-making quality, transparency, and integrity. This is especially significant when there is an urgent need to immediately access or exchange information among healthcare stakeholders such as patients, physicians, and other medical staff. The management of such data in large volumes to support decision-making in a timely manner is a critical requirement in healthcare systems.

2.2 Electromagnetic Interference

While there are several concerns with the use of RFID-based systems in healthcare settings, none of these concerns is as serious as the one associated with electromagnetic interference. In a hospital setting, RFID-based systems are very useful for tracking healthcare equipment, misplacement prevention, and periodic reminders to hospital administration staff as well as triggers when repair or replacement of equipment/devices are necessary. To a certain degree, hospital can be compared to the firm that is composed of a number of divisions or departments. This firm needs to take care of not only people but also thousands of assets. In particular, these assets include movable devices that have to be shared by different divisions across several floors, and include gravity infusion devices (infusion pumps), electrocardiogram (ECG) machines, and ventilators. More uniquely, because life-threatening emergencies are hard to predict, a majority of such devices are extremely crucial in terms of their availability and mobility with respect to caring for the seriously ill patients. Thus, attaching RFID tags to these devices to monitor these life-saving assets in real-time is common in hospital to ensure that they can perform well at the right place and the right time.

However, preliminary testing by the US Food and Drug Administration (FDA) has revealed that the frequency or energy emitted by some RFID-based systems could potentially affect sensitive devices. This issue of electromagnetic interference, that has the potential to affect normal operation of medical devices, could result in serious harm to patients and put their lives in danger (e.g., Ashar and Ferriter 2007, Seidman et al. 2010, Togt et al. 2008). For example, once they are in the proximity of an RFID tag, these critical healthcare equipment may malfunction. The possible consequences include the switch-off for external peacemaker, the disturbance in the atrial and ventricular electrogram curve generated by the pacemaker, change in set ventilation rate of mechanical ventilator, and complete stoppage of syringe pump or renal replacement device. To minimize such EMI-caused danger, hospitals often consider strictly regulating the signal strength used in RFID-based systems.

Similarly, another electromagnetic interference-related challenge in healthcare pertains to avoiding the false read of RFID based systems, because the signal strength is one determining factor in the accuracy of RFID reads. It is certain that the reliable communication between RFID tag and reader is subject

to several conditions, such as the orientation of the tagged object. Moreover, the presence of any liquid, metal, or even mechanical vibration between the tag and its associated reader is not uncommon in hospital environments, but they all can possibly cause coupling interference effect. This effect in turn may reduce the effective reading range and increase the false read rates in RFID-based systems (Haddara and Staaby 2018, Jebali and Kouki 2018). More critically, all of these conditions would become worse, when there is a degradation of the signal strength or sensitivity in RFID based systems due to the electromagnetic interference concern in hospital environments. Therefore, RFID read-rate accuracy is a challenging issue in healthcare settings (e.g., Tu et al. 2009, Zhou and Piramuthu 2018).

2.3 Tag Separation

It is generally assumed that once an RFID tag is placed in position, it stays there forever or at least stays put until its useful life with that object ends and it is then removed from that tagged object. When an RFID tag is placed on top of the tagged item, it is possible for the tag to become separated due to any number of innocuous reasons. However, it is also possible for an adversary to remove the tag from the tagged item in order to achieve some nefarious purpose. While the former results in the 'loss' of the tagged item for all practical purposes in automated systems, the latter could precipitate in serious damage to the patient, healthcare personnel, or healthcare organization. When the former occurs, there has to be a mechanism in place that identifies the event as it occurs and resolves the issue at the earliest possible time before any damage is done. When the latter occurs, it would be nice to have a trigger that goes off immediately so that someone in charge can take necessary action to rectify the situation. The damage caused due to intentional tag separation depends on the motive of the perpetrator and could affect patient(s), hospital personnel, as well as medical equipment and supplies. To our knowledge, the only publication that specifically even considers this eventuality is Tu et al. (2018), who attempt to provide a means to address this situation.

RFID tag detachment in the healthcare domain may result in the management of related supply chain(s) to be compromised. As RFID tags become commonplace to track medical equipment and supplies stock as well as to manage their inventory, the complete trust and reliance placed on RFID tags for these purposes could be violated as tag separation incidents mount. For example, some medicines as well as vaccines are strictly required to be stored under controlled conditions that include allowed temperature range as well as exposure to light. The shelf-life of some of these items are also very short. In RFID-based healthcare systems, the process of checking the stock level and making the appropriate call-offs (orders) to replace anything that is about to run out can be readily automated. The accomplishment of such an automation requires the placement of RFID tags on or inside every medicine bottle, container, tray, etc. in order to allow for ease of access of necessary information such as name, type, vendor, series number, quantity, and expiry date (i.e., medicine profile data). In such

a system that completely relies upon RFID-generated data for a wide range of purposes that include compliance determination to ensuring everyone's safety in a healthcare environment, RFID tag detachments have the potential to lead to unexpected negative consequences. For example, planning based on existing perturbed inventory is bound to be inaccurate due to false read of item profile data that arise from RFID tag detachments. This in turn could result in rush orders or unnecessary inventory. More importantly, when RFID tag detachment instances exceed a critical threshold, related effects on inventory management could prove to be disastrous. For example, the stock level of an important irreplaceable item that generally takes a long time to be restocked could fall short of its safe inventory level. Such an incident could cascade to horrible outcomes.

In a majority of supply chains, the consequences of focal items that are out-of-stock due to tag detachment may be tolerable. However, in a healthcare supply chain in which a life-saving focal item is out-of-stock, RFID tag detachment might result in extraordinary consequences in terms of protection and safety of human lives.

2.4   Relay Attack

RFID-based system are prone to several types of attacks such as cloning, denial of service, eavesdropping, impersonation, malicious code injection, replay attack, side-channel attack, and spoofing (e.g., Mitrokotsa et al. 2010). Over the years, researchers have developed means to address most of these attacks. However, relay attacks have proved to be immune to attempts at addressing them. This is primarily because relay attacks simply relay messages between the two parties (here, RFID tag and reader) without any modification to any of the communicated messages. The adversary mounting a relay attack intends to effectively shorten the physical separation of the tag and reader in order to convince the reader to do something. For example, an adversary can relay messages between an RFID-embedded car key and the car to open and start the car even though the car key might be farther away from the car.

In a healthcare context, an adversary could potentially mount a relay attack on any of the devices that communicate through wireless means and wreak havoc. To our knowledge, so far, none of the medical devices are protected against relay attacks. Although researchers have attempted to develop solutions against relay attacks, none of them are failsafe. A relay attack tricks the tag and reader into believing that they are in communication with each other. A reader and tag can communicate with each other only when they are in close physical proximity of each other. Communication between a tag and a reader are relayed by the adversary during a relay attack. This essentially signifies that the adversary essentially virtually shortens the distance between tag and reader during the attack.

There are two streams of research in this general area. One stream uses distance-bounding method in which the distance between the RFID tag and its reader is estimated by measuring the round-trip time taken by a signal to travel between the two. When the estimated distance is more than the communication

range of the tag and reader, this signifies that a relay attack is in progress. However, since radio waves move at the speed of light, the instruments that measure the signal speed needs to be extremely accurate at the nanosecond scale. Even a small measurement distortion or error could render the measured value invalid. Such a distortion can easily happen since the signal goes from the reader to the tag, which then processes it and then sends the signal back to the reader. If the time it takes for the tag to process the signal is an order more than the signal travel time, the travel time can easily wash away. Clearly, distance-bounding methods face an uphill battle to prove their effectiveness.

The other stream of research uses ambient condition or context at both reader and tag to determine their physical proximity. The underlying idea behind this stream of research is that the ambient conditions at the reader and tag are bound to be the same when they are in close physical proximity of each other. Researchers have used various ambient condition facets such as temperature, light, sound, among others. The issue with this method is that some of the ambient conditions such as light and sound are directional, and could result in different results even though the reader and tag are physically next to each other.

3   Discussion

RFID is a promising technology for healthcare environments where information generated in these tags can be used to replace error-prone human input under some circumstances. Erroneous medical identification, patient misidentification, and such errors in healthcare settings have the potential to do unnecessary harm to patient health. In the United States alone, medical errors have resulted in more than forty million adverse events and have taken away more than two million lives a year (Duroc and Tedjini 2018, Haddara and Staaby 2018). It is therefore not surprising that hospitals and healthcare organizations in general are interested in RFID-based systems to enhance or enable autonomous identification, tracking, monitoring, and management of healthcare procedures and processes that include drug compliance and healthcare supply chain.

Technologies associated with RFID-based systems have been evolving. For example, some passive RFID tags are already capable of measuring physical motion, physical displacement, physical deformation, airflow temperature, and humidity (Tu et al. 2018, Duroc and Tedjini 2018). Meanwhile, novel RFID designs have been focused on leveraging the electromagnetic signature of the tag itself to be the tag identifier (Duroc and Tedjini 2018). This suggests that the unit cost of RFID tag may thus be further reduced because of the removal of the 'chip' inside. There are also novel designs to make RFID tag bendable and combining this with printed sensor or printed battery through 3D printing technology (Duroc and Tedjini 2018, Haddara and Staaby 2018). All of these technological advances have the potential to make the application of RFID tag more attractive and adaptive to healthcare environments. On the other hand, as discussed earlier in this paper, there are still issues for healthcare organizations to seriously consider when adopting RFID-based systems to improve healthcare efficiency and quality.

The concerns with RFID-based systems in the healthcare domain are similar to those in other domains, except for electromagnetic interference and the fact that when something goes wrong, invaluable human lives might be put at stake. Moreover, RFID data privacy and security are critical in healthcare settings. For example, it is a violation of HIPAA (the Health Insurance Portability and Accountability Act) in the United States to let an RFID tag involve any patient data without necessary protection in place. These differences may provide a reasonable explanation as to why the adoption of RFID in the healthcare sector still lags behind that in other industries, despite the many promised benefits of RFID in the healthcare industry. In addition to the challenges that were considered in Kapoor et al. (2009), we identified a few other key issues from the perspective of healthcare applications. More importantly, while all of these issues are non-trivial, prior literature has failed to provide necessary attention to these topics in healthcare settings. We expect that, because of the concise review and critical evaluation of such key issues in this study, there will be more fruitful investigations to address these issues in the near future.

References

Alvarez Lopez, Y., Franssen, J., Alvarez Narciandi, G., Pagnozzi, J., Gonzalez-Pinto Arrillaga, I., Las-Heras Andres, F.: RFID technology for management and tracking: e-health applications. Sensors **18**(8), 2663, 1–17 (2018)

Ashar, B.S., Ferriter, A.: Radio frequency identification technology in health care: benefits and potential risks. J. Am. Med. Assoc. **298**, 2305–2307 (2007)

Chowdhury, B., Khosla, R.: RFID-based hospital real-time patient management system. In: Proceedings of 6th IEEE/ACIS International Conference on Computer and Information Science (ICIS), pp. 363–368 (2007)

Coustasse, A., Arvidson, C., Rutsohn, P.: Pharmaceutical counterfeiting and the RFID technology intervention. J. Hosp. Mark. Public Relat. **20**(2), 100–115 (2010)

Coustasse, A., Tomblin, S.: Impact of radio-frequency identification (RFID) technologies on the hospital supply chain: a literature review. Perspect. Health Inf. Manag. **10**, 1d (2013)

Duroc, Y., Tedjini, S.: RFID: a key technology for humanity. Comptes Rendus Phys. **19**(1–2), 64–71 (2018)

Fazzinga, B., Flesca, S., Masciari, E., Furfaro, F.: Efficient and effective RFID data warehousing. In: Proceedings of the International Database Engineering & Applications Symposium, pp. 251–258 (2009)

Gonzalez, H., Han, J., Li, X., Klabjan, D.: Warehousing and analysis of massive RFID data sets. In: Proceedings of the International Conference on Data Engineering (ICDE) (2006a)

Gonzalez, H., Han, J., Li, X.: Mining compressed commodity workflows from massive RFID data sets. In: Proceedings of the 15th ACM International Conference on Information and Knowledge Management, pp. 162–171 (2006b)

Haddara, M., Staaby, A.: RFID applications and adoptions in healthcare: a review on patient safety. Proc. Comput. Sci. **138**, 80–88 (2018)

Jebali, C., Kouki, A.B.: Read range/rate improvement of an LF RFID-based tracking system. IEEE J. Radio Freq. Identif. **2**(2), 73–79 (2018)

Kapoor, G., Zhou, W., Piramuthu, S.: Challenges associated with RFID implementations in supply chains. Eur. J. Inf. Syst. **18**, 526–533 (2009)

Mathew, P.S., Pillai, A.S., Palade, V.: Applications of IoT in healthcare. In: Sangaiah, A.K., Thangavelu, A., Meenakshi Sundaram, V. (eds.) Cognitive Computing for Big Data Systems Over IoT. LNDECT, vol. 14, pp. 263–288. Springer, Cham (2018). https://doi.org/10.1007/978-3-319-70688-7_11

Meiller, Y., Bureau, S., Zhou, W., Piramuthu, S.: Adaptive knowledge-based system for health care applications with RFID-generated information. Decis. Support Syst. **51**(1), 198–207 (2011)

Mitrokotsa, A., Rieback, M.R., Tanenbaum, A.S.: Classifying RFID attacks and defenses. Inf. Syst. Front. **12**(5), 491–505 (2010)

Mongan, W., et al.: A multi-disciplinary framework for continuous biomedical monitoring using low-power passive RFID-based wireless wearable sensors. In: IEEE International Conference on Smart Computing (SMARTCOMP), pp. 1–6 (2016)

Piramuthu, S., Wochner, S., Grunow, M.: Should retail stores also RFID-tag the 'cheap' item? Eur. J. Oper. Res. **233**(1), 281–291 (2014)

Seidman, S.J., et al.: In vitro tests reveal sample radio frequency identification readers inducing clinically significant electromagnetic interference to implantable pacemakers and implantable cardioverter-defibrillators. Heart Rhythm **7**(1), 99–107 (2010)

Thuemmler, C., Buchanan, W., Fekri, A.H., Lawson, A.: Radio frequency identification (RFID) in pervasive healthcare. Int. J. Healthc. Technol. Manag. **10**(1/2), 119–131 (2009)

Togt, R., van Lieshout, E.J., Hensbroek, R., Beinat, E., Binnekade, J.M., Bakker, P.J.: Electromagnetic interference from radio frequency identification inducing potentially hazardous incidents in critical care medical equipment. J. Am. Med. Assoc. **299**, 2884–2890 (2008)

Tu, Y.-J., Zhou, W., Piramuthu, S.: Identifying RFID-embedded objects in pervasive healthcare applications. Decis. Support Syst. **46**(2), 586–593 (2009)

Tu, Y.-J., Zhou, W., Piramuthu, S.: A novel means to address RFID tag/item separation in supply chains. Decis. Support Syst. **115**, 13–23 (2018)

Williams, T.L., Tung, D.K., Steelman, V.M., Chang, P.K., Szekendi, M.K.: Retained surgical sponges: findings from incident reports and a cost-benefit analysis of radiofrequency technology. J. Am. Coll. Surg. 1–16 (2014)

Zhou, W.: RFID and item-level information visibility. Eur. J. Oper. Res. **198**(1), 252–258 (2009)

Zhou, W., Piramuthu, S.: IoT security perspective of a flexible healthcare supply chain. Inf. Technol. Manage. **19**(3), 141–153 (2018)

Author Index

Printed in the United States
By Bookmasters